First Responder Inspector

Principles and Practice

Jones & Bartlett Learning
World Headquarters
25 Mall Road
Burlington, MA 01803
978-443-5000
info@jblearning.com
www.jblearning.com
www.psglearning.com

National Fire Protection Association
1 Batterymarch Park
Quincy, MA 02169-7471
www.NFPA.org

International Association of Fire Chiefs
8251 Greensboro Drive, Suite 650
McLean, VA 22102
www.IAFC.org

Jones & Bartlett Learning books and products are available through most bookstores and online booksellers. To contact Jones & Bartlett Learning directly, call 800-832-0034, fax 978-443-8000, or visit our website, www.jblearning.com.

Substantial discounts on bulk quantities of Jones & Bartlett Learning publications are available to corporations, professional associations, and other qualified organizations. For details and specific discount information, contact the special sales department at Jones & Bartlett Learning via the above contact information or send an email to specialsales@jblearning.com.

30713-9

Production Credits
Vice President, Innovative Learning and Assessment Solutions: Ada Woo
Senior Director, Content Production and Delivery: Christine Emerton
Director, Product Management: Cathy Esperti
Product Manager: Janet Maker
Manager, Content Development: Tiffany Sliter
Content Manager: Alex Belloli
Content Coordinator: Michaela MacQuarrie
Manager, Intellectual Properties and Content Production: Kristen Rogers
Content Production Manager: Belinda Thresher
Content Production Manager, Navigate: Michael Lepera
Senior Intellectual Property Specialist: Colleen Lamy
Senior Product Marketing Manager: Elaine Riordan
Director, Product Fulfillment: Aaron McKinzie
Purchasing Manager: Wendy Kilborn
Composition: S4Carlisle Publishing Services
Project Management: S4Carlisle Publishing Services
Cover Design: MPS Limited
Text Design: S4Carlisle Publishing Services
Intellectual Property Specialist: Robin Silverman
Intellectual Property Specialist: Faith Brosnan
Cover Image (Title Page and Chapter Opener): © Peter Cutrer
Printing and Binding: Sheridan Michigan

Library of Congress Cataloging-in-Publication Data

Library of Congress Control Number: 2025930392

6048

Printed in the United States of America
29 28 27 26 25   10 9 8 7 6 5 4 3 2 1

Brief Contents

Contents

Acknowledgments

First Responder Inspector: Principles and Practice was developed from content included in ***Fire Inspector, First Edition Revised.*** The content was adapted to cover the knowledge and skills objectives from NFPA 1030 Standard for Professional Qualifications for Fire Prevention Program Positions, 2024 Edition that includes Chapter 6: First Responder Inspector (NFPA 1031).

Special Thanks

A special thank you to Peter Cutrer, CFPS, CFPE, CFI-II, CFI-IAAI and Principal at Virtual Fire Academy and 7CS Consulting LLC for his contributions.

This edition of ***First Responder Inspector*** is published in memory of Fire Chief Raymond Parent, who passed away on April 10, 2024. Chief Parent was a long-time advocate of fire prevention, worked with people to achieve fire safety, and taught the importance of using common sense. You are missed, Chief.

Editorial Board

Shawn Kelley (IAFC)
Director of Strategic Services/GPSS
International Association of Fire Chiefs
Fairfax, Virginia

Ken Holland
National Fire Protection Association
Quincy, Massachusetts

William F. Jenaway
Executive Vice President of Volunteer Firemen's Insurance Services, Inc. (VFIS)
King of Prussia, Pennsylvania

Robert Morris
Director, Fire Prevention Bureau
Darien-Woodridge Fire District
Darien, Illinois

Lead Author

Dr. William F. Jenaway, CFO, CFPS, CSP, CHCM, CPP, MIFE, has over 40 years of fire service and fire inspection experience and holds AS, BS, MA, and PhD degrees. Dr. Jenaway began his career as a volunteer firefighter in East Bethlehem Township in Washington County, Pennsylvania, where he ultimately served as the Fire Chief and Fire Marshal. He subsequently served as the Fire Chief and President of the King of Prussia Volunteer Fire Company in Montgomery County, Pennsylvania. He also served for 15 years as the Chairman of the municipality's Fire and Rescue Services Board.

In 2001, Dr. Jenaway was named "Volunteer Fire Chief of the Year" by *Fire Chief Magazine* at the annual conference of the International Association of Fire Chiefs (IAFC). In 2009, he was elected as Township Supervisor in Upper Merion Township, Pennsylvania, where he served as the liaison to the Planning Commission and the Fire and Rescue Services. In 2011, he served as the Vice Chairman of Upper Merion Township, Pennsylvania.

Dr. Jenaway's over 30-year professional career focused on safety and fire protection engineering. He first served as an insurance field engineer and branch manager, where he was responsible for the inspections of properties, training clients in various loss prevention issues, and testing fire and safety equipment and systems. Dr. Jenaway later served as the Director of Training for the loss control department of CIGNA Property Casualty insurance companies, followed by service as their Vice President of Technical Services. He also served as the Senior Vice President of Risk Control for the Reliance Insurance Companies. In all of these roles, fire and life safety inspection practices were a daily part of his activities.

In addition, Dr. Jenaway served as the Chairman of the Pennsylvania Senate Resolution 60 Commission, which studied the fire and emergency medical service (EMS) delivery system in Pennsylvania and developed legislative and operational recommendations to improve the fire and EMS delivery system. He also served as the Chairman for two National Fire Protection Association (NFPA) Standards: NFPA 1250: *Recommended Practice in Emergency Service Organization Risk Management* and NFPA 1201: *Standard for Providing Emergency Services to the Public.* He has also testified numerous times before local, state, and federal legislative bodies.

Currently, Dr. Jenaway sits on the Underwriters Laboratories Casualty Council. He is a member of the distinguished "Gilmore Commission," appointed by President Clinton and President Bush, which is studying the readiness of the United States to deal with domestic terrorism involving weapons of mass destruction. He is also the Vice Chairman on the Commission on Fire Accreditation International (CFAI).

Dr. Jenaway is the author of more than 200 articles on fire, safety, and management-related topics. He has also written six texts on fire and safety discipline and four children's books. For many years, Dr. Jenaway authored the monthly "Inspect-O-Gram" series for the International Society of Fire Service Instructors (ISFSI) and taught classes for the ISFSI on fire inspection practices.

Dr. Jenaway is currently employed as the Executive Vice President of VFIS. Dr. Jenaway is an adjunct professor in the Public Safety Graduate School at St. Joseph's University in Philadelphia, Pennsylvania, in the Legal Studies Graduate School at California University of Pennsylvania, and in the Fire Science Program at Columbia Southern University, an online university.

Authors

Tim Capehart
Fire Engineer, Retired
Bakersville Fire Department
Bakersville, California

Riley Caton
Gresham Fire and Emergency Services
Gresham, Oregon

Robert M. Coleman, EFO
Chief of Department, Retired
North Attleborough Fire Department
North Attleborough, Massachusetts
Bristol Community College/Adjunct Instructor
Fall River, Massachusetts

Arthur E. Cote, PE, FSPE
Chief Engineer, Retired
National Fire Protection Association
Quincy, Massachusetts

Bradford T. Cronin, CFPS
Harbor Fire Protection
Providence, Rhode Island

Robert Drennen
Director of Emergency Services
Upper Moreland Township
Willow Grove, Pennsylvania
Battalion Chief, Retired
Philadelphia Fire Department
Philadelphia, Pennsylvania

Dr. Robert S. Fleming, EdD, CFPS, CFO, EFO
Professor of Management
Rowan University
Glassboro, New Jersey

William Galloway
Assistant State Fire Marshal
Office of State Fire Marshal
Columbia, South Carolina

Daniel B. C. Gardiner, CFPS
Fire Chief, Retired
Fairfield Fire Department
Fairfield, Connecticut

Jim Goodloe, BA, EFO
Florida Division of State Fire Marshal
Tallahassee, Florida

David A. Hall
Fire Chief
Springfield Fire Department
Springfield, Missouri

Mike Montgomery
Fire Marshal
Humble, Texas

Wm. E. "Bill" Moran
Central Florida Fire Academy
Orlando, Florida

Robert P. Morris
Fire Prevention Bureau
Darien-Woodridge Fire District
Darien, Illinois

Peter J. Mulvihill, PE
North Lake Tahoe Fire Protection District
Incline Village/Crystal Bay, Nevada

Jon Nisja
Fire Safety Supervisor
Minnesota State Fire Marshal Division
St. Paul, Minnesota

Reviewers

Chris Chadwick
Shreveport Fire Academy
Shreveport, Louisiana

Michael D. Chiaramonte
National Fire Academy
Clayton, North Carolina

Richard Connelly
Boston Fire Department
Boston, Massachusetts

Patrick G. Collier
Orland Fire Protection District
Orlando, Florida

Jason D'Eliso
Scottsdale Fire Department
Scottsdale Community College
Scottsdale, Arizona

Rick Fultz
Fresno Fire Department
Fresno, California

Wayne Hamilton
City of Asheville
Asheville, North Carolina

Chris Hofmann
Northland Community and Technical College
East Grand Forks, Minnesota

Jim Iammatteo
Mesabi Range Community and Technical College
Virginia, Minnesota

Mike Julazadeh
Charleston Fire Department
Charleston, South Carolina

Frederick J. Knipper
Duke University/Duke Health System Fire & Life Safety Division
Sanford, North Carolina

Paul A. Paquette
Somerset Fire Department
Somerset, Massachusetts

Gerald H. Phipps, II
Wyoming Fire Academy – WY Department of Fire Prevention and Electrical Safety
Riverton, Wyoming

Rudy Ruiz
Sandusky Career Center
Sandusky, Ohio

Edsel Smith, Jr
West Virginia University Fire Academy and Regional Education Services Agency
Buckhannon, West Virginia

James Smith
City of Kinston
Kinston, North Carolina

Tim Swaim
Kernersville Fire Rescue
Kernersville, North Carolina

Robert Swiger
Raleigh Fire Department
Raleigh, North Carolina

David Telban
Cleveland Fire Department
Cuyahoga Community College
Cleveland, Ohio

Stephen Thomas
Guilford Technical Community College
Jamestown, North Carolina

William Trisler
Commission on Fire Prevention and Control, Connecticut Fire Academy
Windsor Locks, Connecticut

Photographic Contributors

We would like to extend a huge "thank-you" to Glen E. Ellman, the photographer for this project. Glen is a commercial photographer and firefighter based in Fort Worth, Texas. His expertise and professionalism are unmatched!

We would also like to thank Fire Marshal Landon Stallings who opened up his Fire Prevention Bureau, Fort Worth Fire Department, for many of the photographs in this textbook. Specifically we would like to acknowledge the following Fire Inspectors:

Marshal Allen
Fire Captain – Commercial Inspection

Gwen Barnes
Administrative Assistant to Fire Marshal

Tony Blythe
Fire Captain – Addressing

Robert Broadwater
Fire Engineer – Technical Inspections

Debra Clarke
Admin Tech – Revenue

Robert Creed
Fire Lieutenant – Addressing

Brian Hannah
Fire Captain – Technical Inspections

James Horton
Fire Lieutenant – Hazmat

Don Isaacs
Building Permits – Customer Services

Bob Jenkins
Fire Engineer – Technical Inspections

Bob Morgan
Sr. Protection Engineer

Greg Nelson
Fire Protection Engineer

Chip Paiboon
Fire Protection Specialist

Liezel Surel
Admin Tech – Permits

CHAPTER

1

Introduction to First Responder Inspector

NFPA 1030 THAT INCLUDES CHAPTER 6: FIRST RESPONDER INSPECTOR (NFPA 1031)

- 6.2
- 6.3 (pp 4–9)
- 6.3.4 (p 8)
- 6.4
- 6.4.1
- 6.4.2

ADDITIONAL NFPA STANDARDS

- **NFPA 1** *Fire Code*
- **NFPA 101** *Life Safety Code*
- **NFPA 1033** *Standard for Professional Qualifications for Fire Investigator*

KNOWLEDGE OBJECTIVES

After studying this chapter, you will be able to:

1. Describe the functions of the fire inspection unit.
2. Describe the objectives of a fire inspection.
3. Describe the role of the first responder inspector within the fire service.
4. Describe the roles and responsibilities of the first responder inspector.
5. Describe the relationship between state statutes, local ordinances, and fire safety codes.
6. Describe the range of the first responder inspector's legal authority.
7. Describe the types of permits that may be issued to an occupancy.
8. Describe the ethical practices of the first responder inspector.

SKILLS OBJECTIVES

There are no skills objectives for this chapter.

You Are the First Responder Inspector

You have been making joint inspections with a seasoned fire inspector for six months. Most of your inspections have been of restaurants, bars, taverns, and night clubs, with an occasional auto repair shop. You now have been given your first assignment. You are assigned to inspect the local community college, which consists of numerous buildings including the administration building, library, dormitories, classrooms, dining hall, student center, and gymnasium.

1. What is the first priority of fire inspection?
2. Which codes and standards should you review prior to the fire inspection?

Introduction

Effective fire inspections, legal proceedings, and determining the readiness of fire protection systems conducted by the **first responder inspector** can prevent fires and the loss of life. Just as crime prevention has assisted police forces to perform more effectively, fire prevention aids all aspects of the fire service. If fires are prevented from starting, then the true duty of the fire service to the public, protecting lives and property, is accomplished. The first responder inspector provides a valuable service to both the municipality and the property owner by assisting in carrying out the fire department's job of protecting life and property (**FIGURE 1-1**).

The fire inspection unit may be part of the fire department or an independent unit. The fire inspection unit typically focuses on the "Five E's of Fire Prevention," which involve:

FIGURE 1-1 The first responder inspector provides a valuable service to both the municipality and the property owner by assisting in carrying out the fire department's job of protecting life and property.

- Engineering to eliminate hazards through design, correcting hazards, or reducing hazards.
- Educating the public to develop a positive attitude toward fire safety and educating the public on fire prevention codes and regulations.
- Enforcing the codes, standards, and regulations that are designed to help prevent fires, protect occupants, and limit spread.
- Economic incentive, utilizing items such as fire sprinkler systems can actually save a builder money and limit loss. Economic incentive can also be through fines or penalties, and the reduction of such.
- Emergency response to provide the public a quick response in managing emergencies, as well as making sure the responders have the training, education, and equipment necessary to safely do so.

What Is a Fire Inspection?

A **fire inspection** is a visual inspection of a building and its property that is conducted to determine if the building complies with all pertinent statutes and regulations of the jurisdiction (**FIGURE 1-2**). Fire inspections are conducted to reduce the risk of fire and maintain a reasonable level of protection of life and property from the hazards created by fire, explosion, and hazardous materials. Fire inspections are conducted at businesses, apartment buildings, schools, hospitals, places of public assembly, and other occupancies with the exception of one- or two-family dwellings. Safety surveys may be conducted in homes at the request of the owner, but these are typically done simply as a courtesy.

A fire inspection may include inspecting fire suppression, fire alarm, and detection systems; conducting occupant evacuation drills, and providing guidelines to the

FIGURE 1-2 A fire inspection is a visual inspection of a building and its property that is conducted to determine if the building complies with all pertinent statutes and regulations of the jurisdiction.

building owners on fire safe materials. In addition to standard fire inspections, there are also general fire safety or process inspections that are performed to check for overcrowding, blocked exits during business hours, or other types of unsafe practices that may be occurring while the building or space is occupied, placing the lives of the occupants at risk. Examples include a new movie release at the local movie theatre resulting in larger than normal crowds, or overflowing storage in mercantile properties during the holiday season. General fire safety or process inspections may be the result of a complaint, such as overcrowding in a bar.

Fire Inspection and the Fire Service

Depending upon the community's size, complexity, state and local legislative requirements, and multiple other factors, fire inspectors may work within the community's fire department, the community's independent code enforcement agency, or in a business that performs fire inspections as a contracted service by the community. For many years, the responsibilities of fire inspection lay strictly within the fire department; however, the complexities of the task, coupled with legislative, training, and mandated services resulted in some communities creating an independent agency or contracting the service to a private enterprise. In some communities, local fire department engine or ladder companies continue to conduct fire inspections in addition to pre-planning inspections. In the end, the local government determines where and who performs fire inspections in the community.

PRO TIPS

Often the term fire safety inspection is used to describe the activities being performed during the fire inspection. The term is also regularly applied to inspections that involve the enforcement of the NFPA 101, *Life Safety Code*, which expands the scope of the inspection to include general safety issues. The terms fire inspection and fire safety inspection may be used separately or interchangeably.

Additional Fire Prevention Roles within a Fire Department

If a fire inspector works within a fire department, the fire inspector may hold multiple roles during a profession, particularly in smaller fire departments. Some of the more common positions that fire inspectors may also assume include:

- **Fire marshal**: Fire marshals inspect businesses and enforce public safety laws and fire codes. They may respond to fire scenes to help investigate the cause of a fire.
- **Fire inspector**: Fire inspectors use and apply codes and standards while performing fire inspections, plans reviews, facilitates training, and resolves code-related issues.
- **Fire investigator**: Fire investigators may have full police powers to investigate and arrest suspected arsonists and people causing false alarms.
- **Fire and life safety educator**: These individuals educate the public about fire safety and injury prevention and present juvenile fire safety programs.
- **Fire protection engineer**: The fire protection engineer usually has an engineering degree, reviews plans, and works with building owners to ensure that their fire suppression and detection systems will meet code and function as needed. Some fire protection engineers actually design these systems.

Roles and Responsibilities for the First Responder Inspector

According to NFPA, *1030, Standard for Professional Qualifications for Fire Prevention Program Positions* a first responder inspector conducts basic fire inspections and applies codes and standards (**FIGURE 1-3**). The First Responder Inspector must also meet the job

FIGURE 1-3 A hearing is held to discuss the new or modified code by the elected officials with the general public.

performance requirements specified in NFPA 1030, including:

- Inspect structures in the field and write reports based on observations and findings.
- Identify the need for a permit and communicate the requirements for such.
- Investigate assigned complaints, record the information and findings, and forward them to the Authority Having Jurisdiction (AHJ).
- When presented with a fire protection, fire prevention, or life safety issue, identify the applicable code, standard, or policy that is being violated.
- Participate in legal proceedings and provide testimony or written comments as required with factual accuracy.
- Identify the occupancy classification of a single-use occupancy.
- Verify the posted occupant load to a specific occupancy classification so that the building or structure is within compliance.
- Verify that the means of egress elements are maintained and readily accessible, unlocked, and free of obstructions.
- Determine if existing fixed fire suppression systems are working properly through observation.
- Determine if existing fire detection and alarm systems are working properly through observation.
- Determine if existing fixed portable fire extinguishers are working properly through observation.
- Inspect the emergency access for the fire department to an existing site.
- Verify that hazardous materials, flammable and combustible liquids, and gases are identified, documented, and reported in accordance with local codes and laws.
- Recognize a hazardous fire growth potential in a building or space.
- Determine code compliance, given the codes, standards, and policies of the jurisdiction and a fire protection issue.

Codes and Standards

Codes and standards are regulatory or industry-adopted documents that provide guidance for the safe construction and use of buildings. A **code** is a rule or law established for enforcement of a fire protection, life safety, building construction, or property maintenance issue. A code can be independently incorporated into law without supporting documents or standards. Codes relate to many types of equipment used in construction, such as heating, ventilating, and air- conditioning equipment; construction materials; and similar products. A **standard** is a document in which are specifications or requirements defining the minimum levels of performance, protection, or construction. The standard may or may not be a legally mandated document, but considered an industry best practice to follow.

PRO TIPS

A code tells you WHEN something is required; a standard tells you HOW to accomplish it.

Model Code Organizations

Model code organizations write or develop codes and standards for incorporation by governmental and other regulatory agencies such as insurance organizations. Model code organizations include the National Fire Protection Association (NFPA), International Code Council (ICC), and the American Society of Mechanical Engineers (ASME). To ensure a high level of acceptance of codes, model code organizations adopt stringent guidelines for developing codes. Technical advisory committees are made up of the different interest groups that will be directly affected by the codes. For example, a sprinkler manufacturer may sit on the committee revising the code on fire suppression systems. A fire code technical advisory committee may include representatives from the fire service, the insurance industry, equipment manufacturers, and contractors. The various members of the technical advisory committees provide a balance of perspectives and views in the development of the code, assuring that all users are involved in the decision making process.

Prescriptive Codes and Performance-Based Codes

Prescriptive codes and performance-based codes are used during the fire inspection process. A **prescriptive code** lists the specific details the installation or construction must meet, such as the type of electrical wiring to be used in an occupancy. A **performance-based code** outlines a requirement that a design has to meet, but does not state that a particular method or material must be used to meet the requirement. An example of a performance-based code is the requirement that a fire pump serving a high-rise building must be protected by 2-hour fire rated construction or be physically separated by 50 ft from the building. The first responder inspector uses both types of codes to ensure that a building is safe to be occupied.

Code and Standard Incorporation

Local, county, district, or state officials may select for incorporation the codes or standards that meet the specific needs of the jurisdiction. The code or standard incorporation process typically follows this path:

- The need to update to a new code or incorporate by referencing an existing code arises when a new code is created by a model code agency, a fire inspector expresses the need for a code to enforce, or a code needs modification.
- The proposed modified language is prepared by technically competent staff or by consultants or the new code is collected.
- The proposed modified code or new code is provided to elected officials or an administrative committee or panel for review.
- Upon review and any corrections, the modified or new code are posted for review by the general public.
- A hearing is held to discuss the new or modified code by the elected officials with the general public (Figure 1-3).
- Input is provided, considerations for changes are taken, modifications may be made, and the elected officials then vote the code into statute.
- There may be an appeals process depending on the level of government enacting the regulations, state law governing incorporation of safety regulations, and related oversight agencies such as State Fire Marshal's office, Public Utilities Commission, etc.

As a rule, codes and standards are incorporated under most state laws or statutes or by local ordinances. State statutes will generally provide guidance on how to apply codes and standards in your local jurisdiction. However, there are some federal laws governing fire safety issues. For example, the 1990 *Hotel and Motel Safety Act* requires that federal government employees traveling on official business must stay in hospitality properties equipped with a fire sprinkler system.

PRO TIPS

An effective first responder inspector is trained to recognize fire hazards and to correct problem situations. You also must be capable of communicating information to building occupants and owners both orally and in written form. As a first responder inspector, you must be able to translate codes and standards into language that the layperson can understand in order to explain to property owners why certain conditions present a fire hazard and how to correct the situation. For example, you may be asked to explain the difference between a smoke detector and carbon monoxide detector to a homeowner, and why both may be needed.

PRO TIPS

Recommended practices or guides are manufacturer or industry developed techniques to perform an activity in a safe manner. They are typically not incorporated into a code, ordinance, or law, but are simply the best practice to follow.

Understanding the Legal Processes

The legal authority for fire inspection varies by state to state and, in many cases, varies within each state. The job description of the first responder inspector should list the limits of legal authority that the position holds in the local jurisdiction, and how to access or elevate the issue to the next level of expertise or authority when needed. As part of the introductory training process for the first responder inspector, a specific session must be included on the authority of the first responder inspector and the process that must be used when citations are required or when legal action must be taken upon the occupant or owner as the process will vary by community.

Authority Having Jurisdiction

Who is the Authority Having Jurisdiction (AHJ) for fire inspections? According to the NFPA, the AHJ is the organization, office, or individual responsible for enforcing the requirements of a code or standard, or for approving equipment, materials, an installation, or a procedure. The AHJ receives its authority through local or state government. For example, in Florida, the AHJ for fire inspections is the local fire chief of the fire department. If there is not a local fire department, then a government official is designated by the local government. The Office of State Fire Marshal has direct responsibility for all state-owned facilities. However, authority is granted to the State Fire Marshal to inspect other occupancies when there is reason to believe that a fire safety violation exists, if a local fire inspector is unavailable, or if a higher level of expertise or authority is necessary.

The authority to perform fire inspections differs from state to state. Some states require state oversight of specific occupancies such as nursing homes, daycare centers, and certain personal care facilities. You must be familiar with the laws and regulations of your state and local government prior to engaging in fire inspection and code enforcement activities.

State and Local Law

State and local law affects every aspect of your job. For example, your fire inspection unit may have specific requirements that every written notice must follow. In order to avoid legal disputes, a standardized notice of violation that has been reviewed for legal sufficiency by the attorney for the jurisdiction should be used by all personnel.

Range of Authority

The range of authority of the first responder inspector is defined by state or local law. The range of authority typically allows for greater authority for a fire inspector. The range of authority may deal with types of properties that the first responder inspector can inspect, the size of properties that the first responder inspector may inspect, the types of inspection that the first responder inspector may perform, for example, hazard-based or fire-protection-equipment-based, and the enforcement authority of the first responder inspector versus the fire inspector.

Legal Proceedings

As a first responder inspector, you may be involved in legal proceedings. The results of a fire inspection, failure by a building owner to comply with corrective actions, and resulting fires may require you to appear in court. These legal proceedings may be criminal or civil in nature. Legal proceedings may require you to present documentation of your training and expertise, the local procedures used in conducting fire inspections in your jurisdiction, and certification to support any information you provide as testimony.

Permits

Many fire inspection activities result in the issuance of a **permit**. A fire inspection may be a step in the permit application process, depending on the type of permit and local procedures. A permit is a document stating the compliance with or restrictions for use based upon applicable codes. The permit may be a:

- Use and occupancy permit—A permit to occupy a structure. For example, a nail salon in a mall.
- Special use permit—A permit to perform specified activities for a specified time period. For example, a carnival or outdoor fair.
- Fire protection equipment permit—A permit to install or use a particular fire protection equipment device. For example, a permit to install a sprinkler system in a storeroom of a shoe store or to use a fire hydrant.
- Hazardous material use permit—A permit to use a specific product for a specific purpose for a specific time period. For example, a permit to use a chemical during a manufacturing process.

Ethics and the First Responder Inspector

Ethics are critical in fire inspection. An first responder inspector must not be influenced by business or political interests. Inappropriate behavior by individuals in power or those holding the public trust gets frequent attention by the media. The fire service is not exempt. Most of the time, first responder inspectors make ethical decisions; however, some make unethical choices that often appear in the newspaper and have very negative consequences for the individual, the organization and the Fire Service in general.

Ethical choices are based on a value system. The first responder inspector has to consider each situation, often subconsciously, and make a decision based on his or her values. If the organization has clear values that are part of a strong organizational culture, the first responder inspector uses the organization's value system. If the values are not clear, the individual substitutes his or her own value system.

The key to improving ethical choices is to have clear organizational values. This can be accomplished by:

- Having a code of ethics that is well known throughout the organization
- Selecting employees who share the values of the organization
- Ensuring that top management exhibit ethical behavior

PRO TIPS

"Honesty + Integrity = Credibility. If you lose one of the first two, you won't have the third, which is the most important qualification you can have in this field."—Chief Raymond Parent

- Having clear job goals
- Having performance appraisals that reward ethical behavior
- Implementing an ethics training program

Even at the company level, these values can be implemented to help prevent undesirable ethical choices. One way to help judge a decision is to ask yourself three questions:

- What would my parents and friends say if they knew?
- Would I mind if the paper ran it as a headline story?
- How does it make me feel about myself?

Asking these questions can help prevent an event that could devastate the department's and the first responder inspector's reputation for years to come.

Career Development

The first responder inspector in today's world is expected to continually develop his or her professional skills and knowledge. The required skills and knowledge are defined not only in NFPA standards but also various state laws. Professional organizations or state agencies offer continuing education and professional credentialing. Depending upon the state, first responder inspectors may be required to complete certifications or tests to document performance capabilities.

WRAP-UP

CHAPTER SUMMARY

- A fire inspection is a visual inspection of a building and its property that is conducted to determine if the building complies with all pertinent fire and safety statutes and regulations of the jurisdiction.
- Fire inspections are conducted to reduce the risk of fire and maintain a reasonable level of protection of life and property from the hazards created by fire, explosion, and hazardous materials.
- Depending upon the size, complexity, state and local legislative requirements, and multiple other factors, first responder inspectors may work within the community's fire department, the community's independent code enforcement agency, or in a business that performs fire inspections as a contracted service by the community.
- Codes and standards are regulatory or industry incorporated documents that provide the minimum requirements for the safe construction and use of buildings.
- A code is a rule or law established for enforcement of a fire protection, life safety, building construction, or property maintenance issue. Codes relate to many types of equipment used in construction, such as heating ventilating and air-conditioning equipment, construction materials and similar products.
- A standard is a document containing specifications or requirements defining the minimum levels of performance, protection, or construction. The standard may or may not be a legally mandated document, but is considered an industry best practice to follow.
- The job description of the first responder inspectors should list the limits of legal authority that the position holds in the local jurisdiction, and how to gain the next level of expertise or authority when needed.
- Many fire inspection activities result in the issuance of a permit. A permit is a document stating the compliance with or restrictions for use based upon applicable codes.

KEY TERMS

Code A standard that is an extensive compilation of provisions covering broad subject matter or that is suitable for incorporation into law independently of other codes and standards.

Fire and life safety educator A member of the fire department who deals with the public on education, fire safety, and juvenile fire safety programs.

KEY TERMS CONTINUED

Fire inspection A visual inspection of a building and its property to determine if the building complies with all pertinent statutes and regulations of the jurisdiction.

First responder inspector An individual at the first level of progression who has met the job performance requirements specified in this standard for first responder inspector. The first responder inspector conducts basic fire inspections and applies codes and standards. (NFPA 1030)

Fire inspector An individual at the second level of progression who has met the job performance requirements specified in this standard for fire inspector. The fire inspector conducts most types of inspections and interprets applicable codes and standards. (NFPA 1030)

Fire investigator An individual who has demonstrated the skills and knowledge necessary to conduct, coordinate, and complete an investigation. (NFPA 1033)

Fire marshal A member of the fire department who inspects businesses and enforces laws that deal with public safety and fire codes.

Fire protection engineer A member of the fire department who works with building owners to ensure that their fire suppression and detection systems will meet code and function as needed.

Performance-based code Outlines the requirement that a design has to meet, but does not state that a particular method or material must be used to meet the requirement.

Permit A document issued by the authority having jurisdiction for the purpose of authorizing performance of a specified activity. (NFPA 1)

Prescriptive code Defines the specifics of a material of construction or action to be taken; such as the type of electrical wiring to use, based on the anticipated usage or requirement to conduct evacuation drills in a structure.

Standard A document, the main text of which contains only mandatory provisions using the word "shall" to indicate requirements and that is in a form generally suitable for mandatory reference by another standard or code or for adoption into law. Nonmandatory provisions shall be located in an appendix or annex, footnote, or fineprint note and are not to be considered a part of the requirements of a standard.

You Are the First Responder Inspector

A developer wants to build a three-story commercial office building with residential condominiums on the second and third floors. Their first stop is at your office to find out "What do you require?" Their second question is, "What do I have to do as far as plans are concerned?" What the developer is really asking is, "What are the fire protection requirements for the building that is going to be built?" and "How often do we have to check with you during the building phase?" Additionally, the developer will be asking about the coordination between the fire protection requirements and any other code requirements such as the building department or zoning, and the timelines of such compliance efforts.

To formulate an answer, you must know the occupancy type, specific occupancy-based requirements, and materials being used. Once these questions are answered you will be able to give the developer an idea of when you need to see the plans, and at what point you will be visiting the construction site.

1. A code is:
   - **A.** determined by the first responder inspector based on what he/she feels is best for the specific location or occupancy.
   - **B.** a legal requirement, a building feature, or an operating feature at a certain location.
   - **C.** the manner that life safety is accomplished in a building.
   - **D.** a legal mandate allowing the inspector access to buildings.

2. A standard is:
- **A.** how a code requirement is implemented.
- **B.** a set of specifications or requirements that is considered industry best practice, and meets a minimum level of performance guidelines.
- **C.** a customary requirement that is required by all.
- **D.** an everyday way of accomplishing something occupancies of a specific class. regardless of the occupancy.

3. The most important tool the first responder inspector possesses is:
- **A.** the ability to document the inspection.
- **B.** maps and previous inspection reports of the occupancy.
- **C.** support from the building's owner.
- **D.** knowledge.

4. Any non-compliance (code violation) that has been noticed during the fire inspection:
- **A.** must be documented in the written report of the inspection.
- **B.** must be told to the person in authority of the occupancy during the closing conference.
- **C.** may be overlooked by the first responder inspector if there is good reason at the time.
- **D.** shall be shared with the personnel at the nearest fire station for their safety.

CHAPTER

2

Types of Occupancies

NFPA 1030 THAT INCLUDES CHAPTER 6: FIRST RESPONDER INSPECTOR (NFPA 1031)

First Responder Inspector

- 6.5.3
- 6.5.5 (pp 43–52)

ADDITIONAL NFPA STANDARDS

- **NFPA 101** *Life Safety Code*
- **NFPA 5000** *Building Construction and Safety Code, or the International Building Code*

KNOWLEDGE OBJECTIVES

After studying this chapter, you will be able to:

1. Describe how the occupancy classification of a building is determined.
2. List the fifteen specific occupancy groupings.
3. Describe the identifying characteristics of one- and two-family dwellings.
4. Describe the identifying characteristics of lodging or rooming houses.
5. Describe the identifying characteristics of hotels and dormitories.
6. Describe the identifying characteristics of apartment buildings.
7. Describe the identifying characteristics of residential board and care occupancies.
8. Describe the identifying characteristics of healthcare occupancies.
9. Describe the identifying characteristics of ambulatory healthcare occupancies.
10. Describe the identifying characteristics of day-care occupancies.
11. Describe the identifying characteristics of educational occupancies.
12. Describe the identifying characteristics of business occupancies.
13. Describe the identifying characteristics of industrial occupancies.
14. Describe the identifying characteristics of mercantile occupancies.
15. Describe the identifying characteristics of storage occupancies.
16. Describe the identifying characteristics of assembly occupancies.
17. Describe the identifying characteristics of detention and correctional occupancies.

SKILLS OBJECTIVES

There are no skills objectives for this chapter.

You Are the First Responder Inspector

You are preparing to inspect a recently renovated historic structure in your town's center. In the past, this two-story Victorian was used as a private residence and fell into dis-repair. You read in the newspaper that a local history museum purchased the property and renovated it, turning it into a museum. The first thing you must determine is what the current occupancy classification is, as no doubt the code requirements for an old single-family residence are much different than its current use. Before knowing what section of the code you should familiarize yourself with in order to make a proper inspection, you visit the building and observe its daily operations.

1. Based on what you know about the building, what type of occupancy do you think it is?
2. What specific features would be necessary to change an occupancy from a one- or two-family dwelling to an assembly occupancy?

Introduction

In the model building codes, the term **occupancy** refers to the intended use of a building. The occupancy classification of a building is determined by the current use of that building. A building that was constructed and used for one purpose, such as an elementary school, may be later used for a different purpose, such as a community center. Prior to conducting an inspection, the first responder inspector must determine the previously approved occupancy classification and verify whether the building is currently being used within the appropriate occupancy classification. Because a building can change occupancy classifications over time, it is critical that you correctly identify the current use of the building and the appropriate occupancy classification under the applicable building codes and regulations.

Codes and regulations will dictate the requirements necessary to conduct the fire inspection accurately and provide standards that you will use to evaluate the structure. Depending upon your jurisdiction, you may enforce NFPA 101, *Life Safety Code*, NFPA 5000, *Building Construction and Safety Code* or the *International Building Code*. Both NFPA 101 and NFPA 5000 are published by the National Fire Protection Association and address construction, protection, and occupancy features necessary to minimize the danger to life and property from the effects of fire. The *International Building Code* is a model building code developed by the International Code Council. While the various model codes in use today vary in terms of their specific requirements, they are fairly similar in terms of how they classify occupancies. However, due to specific differences in model codes, it is critical to become familiar with the specifics of the model code used within your jurisdiction.

Occupancy Classification

It is extremely important that the correct occupancy classification be determined because the model code requirements differ for each type of occupancy. Improperly classifying a building prior to inspecting it can result in an inadequate level of fire and life safety, such as a failure to require necessary fire protection systems or exits. Similarly, an incorrect occupancy classification could result in requiring fire protection systems that are not required for the occupancy and increasing the costs for the owner.

According to the NFPA 101, *Life Safety Code,* there are fifteen specific occupancy groupings:

1. One- and two-family dwellings
2. Lodging or rooming houses
3. Hotels and dormitories
4. Apartment buildings
5. Residential board and care
6. Health care
7. Ambulatory health care
8. Day-care
9. Educational
10. Business
11. Industrial
12. Mercantile
13. Storage
14. Assembly
15. Detention and correctional

PRO TIPS

A dwelling unit is expected to be occupied by members of a single family with not more than three outsiders, if any, accommodated in rented rooms.

One- and Two-Family Dwelling Units

According to NFPA 5000, *Building Construction and Safety Code*, a **one- or two-family dwelling** is a building that contains no more than two dwelling units with independent cooking and bathroom facilities (**FIGURE 2-1**).

The way in which the living units are separated from each other also helps to determine the occupancy classification. For example, a row of townhouses incorporating complete vertical wall separation between the individual units and an independent means of egress from each unit are considered to be separate dwelling units.

The regulations with respect to inspection of these residential units vary by municipality. A jurisdiction may require an occupancy inspection at the time a home is built, sold, or refinanced. Residents may also request these inspections in the interest of fire and life safety.

In conducting an inspection of a one- or two-family home, you will typically focus on the means of escape, the interior finishes, and fire protection. The placement and functionality of fire protection devices and/or systems should be examined, including residential sprinklers if provided and/or required. Hazards such as utilities, alternative heating devices, and storage should also be addressed during the inspection.

FIGURE 2-1 A one- or two-family dwelling unit only contains no more than two independent cooking and bathroom facilities.

As a first responder inspector, you have an essential role as a fire safety educator. You will always want to provide fire safety education when your inspection reveals that it is needed. The focus of these educational activities should be on maintaining working smoke detection systems, creating a predetermined escape plan, and ensuring proper housekeeping. Flammable materials, for example, should not be stored in close proximity to a gas stove or electric range.

PRO TIPS

In a condominium building containing multiple dwelling units, each dwelling unit is considered a private dwelling for entry purposes. An individual unit is usually protected by the Fourth Amendment relative to inspecting the unit without first obtaining the occupants' permission; however, common areas of the condominium building may be inspected at all reasonable times as permitted by code.

Lodging or Rooming Houses

According to NFPA 101, a **lodging or rooming house** is a building or portion thereof, not categorized as a one- or two-family dwelling, that provides sleeping accommodations for a total of 16 or fewer people on a transient or permanent basis (**FIGURE 2-2**). A lodging house may provide meals but should not include separate cooking facilities for individual occupants. Personal care services are not provided in lodging houses. Guest houses, bed and breakfasts, small inns or motels, and foster homes fall under this category. Small sleeping accommodations in other occupancies, such as a fire station, can also fall under this category.

The nature of these occupancies may result in occupants being unaware of their surroundings and thus dependent on the building's fire and life safety features. Of particular concern as you inspect these buildings will be the building compartmentation and the provision of adequate escape routes. The adequacy of the escape routes and its proper signage should be carefully considered during your inspection.

Additional inspection considerations include compliance with use and code requirements, interior finishes, protection of exits, building services, and fire protection

FIGURE 2-2 A lodging house will have room for a maximum of 16 guests.

systems. Inspections should address housekeeping, electrical installations, heating appliances, cooking operations, fire detection and alarm systems, and sprinkler systems. Your goal is to help prevent the occurrence of a fire, as well as ensuring that necessary building construction features and fire protection systems are in place to quickly alert the building occupants in the event of a fire and provide the means to safely exit the building.

Hotels

According to NFPA 101, a **hotel** is a building or group of buildings under the same management that has sleeping accommodations for more than 16 people and is primarily used by transients for lodging, with or without meals (**FIGURE 2-3**). This category includes modern fire-resistant high-rise hotels, motels, inns, and clubs. Hotels present a diverse range of fire and life safety issues that you must be aware of and thoroughly address. In addition to sleeping rooms, a hotel may have restaurants, meeting rooms, ballrooms, stages, theatres, retail stores, garages, offices, storage rooms, and maintenance shops, each of which may fall under a different occupancy class. While the guest rooms would be residential occupancies, the ballrooms, restaurants, and theatres would be assembly occupancies. Offices would be business occupancies, while parking garages and storage areas would be storage occupancies. Retail stores would be mercantile occupancies, and maintenance shops would be industrial occupancies.

FIGURE 2-3 A hotel has room to house more than 16 occupants.

PRO TIPS

Laundry rooms with large industrial washers and dryers are common in hotels. These areas present fire prevention issues, including a large fire load, lint buildup inside and outside of the dryers, linen carts blocking the means of egress, storage, and appliances that have some type of heating element. Ensure that these areas have a daily cleanup and maintenance plan to decrease the likelihood of a fire.

When these various occupancies are separated with their own means of egress, they can be treated as **separate occupancies**. If they share a means of egress, the fire protection requirements are based on the most restrictive requirements of either type of occupancy. If the hotel had a freestanding maintenance shop it would be treated as a separate facility and subject to the requirements of that occupancy class, whereas a maintenance shop housed within the hotel building itself would likely fall under the more restrictive provisions of the other occupancies housed within the building.

All interior areas of the building including guest rooms, top floors, front desk areas, lobbies, assembly rooms, and service areas should be inspected. In this type of occupancy, the building occupants are likely to be unfamiliar with their surroundings and will depend on the building design and fire and life safety protection to ensure their life safety. Fire alarm systems should be clearly heard in all rooms of the building, and clearly marked and lighted exit routes should be available throughout the building.

You should verify that the facility has an appropriate fire and emergency plan and that building personnel are capable of administering its procedures. Personnel should have received the necessary training and practice to know the plan and enact it successfully. In reviewing documentation of staff training and participation in drills, you will want to take into account the impact of staff turnover, which can be fairly significant within the hospitality industry.

Dormitories

According to NFPA 101, a **dormitory** is a building or space in a building in which group sleeping accommodations are provided for more than 16 persons who are not members of the same family. This lodging may be provided in one room or a series of closely associated rooms, under joint occupancy and single management, with or without meals, but without individual cooking facilities (**FIGURE 2-4**). Inspection practices of dormitories focus on fire detection systems and occupancy evacuation and must also address issues that are unique to dormitory living.

The inspection should include means of egress, interior finishes, building contents, housekeeping, and fire protection systems. The use of alternative heating devices and overloading electrical circuits can be a concern. Many jurisdictions, whether at the local or state level, may have specific fire and life safety provisions with respect to college and university dormitories. It is important that you ensure that there is a fire and emergency plan including appropriate escape routes in place and that building occupants and staff have received the necessary fire safety training and have drilled in accordance with the emergency plan.

Apartment Buildings

According to NFPA 101, an **apartment building** is a building or portion thereof containing three or more dwelling units with independent cooking and bathroom facilities (**FIGURE 2-5**). The nature and character of the occupancy may be based on the location and design of the building as well as the age and social status of the occupants. You must be aware of the code requirements that were in place at the time the building was constructed. Based on the adoption of new codes and requirements over time, the code requirements are often different for existing apartment buildings and new apartment buildings, and you must be familiar with these differences when inspecting buildings and enforcing codes and regulations. An example of this could involve the permitted travel distances to exits in an apartment building built in 1940 versus 2010.

When inspecting an apartment building, it is recommended that you begin by conducting a general observation of the exterior of the building. Upon moving to the interior of the building, you should consider the occupant load, dwelling units, means of egress, interior finishes, the protection of openings, waste chutes, and hazardous areas.

Inspection of fire and life safety equipment should include fire alarm systems, portable fire extinguishers, sprinkler and/or standpipe systems, and lighting. The importance of the proper placement and maintenance of smoke alarms should be emphasized during any inspection of an apartment building to both building staff and residents.

FIGURE 2-4 Dormitories provide unique challenges for the first responder inspector.

FIGURE 2-5 Apartment buildings are often the most difficult occupancies to inspect.

PRO TIPS

"Apartment hotels" should be classified as hotels, because they are potentially subject to the same transient occupancy as hotels. Transients are those who occupy accommodations for less than 30 days.

Residential Board and Care Occupancy

NFPA 101 defines a **residential board and care occupancy** as a building or portion thereof that is used for lodging and boarding of four or more residents, not related by blood or marriage to the owners or operators, for the purpose of providing personal care services (**FIGURE 2-6**). These occupancies provide lodging, boarding and personal care services and include residential care homes, personal care homes, assisted living facilities, and group homes.

Examples of residential board and care occupancies include:

- Group housing arrangement for physically or mentally handicapped persons who normally attend school in the community, attend worship in the community, or otherwise use community facilities
- Group housing arrangement for physically or mentally handicapped persons who are undergoing training in preparation for independent living, for paid employment, or for other normal community activities
- Group housing arrangement for the elderly that provides personal care services but that does not provide nursing care

FIGURE 2-6 An example of a residential board and care occupancy.

- Facilities for social rehabilitation, alcoholism, drug abuse, or mental health problems that contain a group housing arrangement and that provide personal care services but do not provide acute care
- Assisted living facilities
- Other group housing arrangements that provide personal care services but not nursing care

You must recognize the difference between personal care and healthcare occupancies. Personal care occupancies include assisting occupants with many of the activities of daily living, such as bathing and dressing. Personal care does not include medical care. If nursing care is provided, then the facility is considered to be a healthcare occupancy; however, if nursing care is not provided, it is considered to be a residential board and care occupancy. It is important that you accurately determine the classification of the facility and corresponding requirements for compliance with relevant codes and/or regulations.

A primary concern in inspecting these occupancies is the capability to evacuate residents in the event of a fire or other emergency. The characteristics of the occupants, means of egress, and egress capacity should be reviewed. Additional considerations when conducting the inspection include building compartmentation, protection of vertical openings, hazardous areas, fire detection and alarm systems, and fire suppression equipment. The occupants of these facilities rely in varying degrees on staff assistance and building fire and life safety provisions. It is imperative that you verify that all fire protection and life safety equipment is in operational condition and has been properly maintained and tested.

Given the role that facility staff personnel play in ensuring the life safety of residents, it is crucial that you validate that an appropriate fire and emergency plan is in place. You must ensure that facility staff and residents are aware of its procedures and that all staff members have received the necessary fire and life safety training. Staff members must also prove their knowledge and skills through fire drills.

Healthcare Occupancy

NFPA 101 defines a **healthcare occupancy** as an occupancy used for purposes of medical or other treatment or care of four or more persons on an inpatient basis where such occupants are mostly in capable of self-preservation due to age, physical or mental disability, or because of security measures not under the occupant's control. These occupants also include infants, convalescents, or infirmed aged persons (**FIGURE 2-7**).

FIGURE 2-7 An example of a healthcare occupancy.

Note that the definition of healthcare occupancy specifies inpatients. A healthcare facility used only for outpatients is addressed as an ambulatory healthcare occupancy.

Included within this occupancy classification are hospitals or other medical institutions, nurseries, nursing homes, and limited care facilities. Because patients within these occupancies are presumed to be incapable of self-preservation in terms of seeking safe refuge should a fire occur, their life safety is based on the incorporation of fire and life safety features into the building and the staff response to a fire or other emergency situation. Life safety features include building construction features such as compartmentalization, the provision of automatic sprinklers, and the training of facility staff in the procedures that should be followed in the event of a fire. These occupancies are often designed to provide for "defending in place" in the event of a fire or other emergency. Under this approach, building features such as compartmentalization and protection of openings are designed to provide protection of patients during a fire.

Essential elements of the inspection of a healthcare occupancy involve ensuring appropriate means of egress in compliance with code and that all fire detection and suppression systems are properly maintained and fully functional. Inspection and maintenance documentation for these systems should be thoroughly reviewed.

PRO TIPS

Typically, healthcare occupancies are INpatient, while ambulatory healthcare occupancies are OUTpatient.

Ambulatory Healthcare Occupancies

An **ambulatory healthcare occupancy** is defined in NFPA 101 as an occupancy used to provide outpatient services or treatment to four or more patients simultaneously. These services include one or more of the following:

- Treatment for patients that renders the patients incapable of taking action for self-preservation under emergency conditions without the assistance of others.
- Anesthesia that renders the patients incapable of taking action for self-preservation under emergency conditions without the assistance of others.
- Emergency or urgent care for patients who, due to the nature of their injury or illness, are incapable of taking action for self-preservation under emergency conditions without the assistance of others (**FIGURE 2-8**).

Facilities falling under this occupancy classification do not provide overnight sleeping accommodations, while those under the healthcare occupancy do. This occupancy class includes freestanding emergency medical units, hemodialysis centers, and outpatient surgical units, often referred to as day surgery centers, where general anesthesia is administered. Oral surgery centers also fall under this occupancy classification. Practices, such as the typical physician's or dentist's office, where four or more patients are *not* rendered incapable of self-preservation are not classified as ambulatory healthcare occupancies. Physician and dental offices are classified as business occupancies.

FIGURE 2-8 An example of an ambulatory healthcare occupancy.

Considerations when inspecting an ambulatory healthcare occupancy include means of egress and protection features such as building construction, smoke barriers, fire detection and alarm systems, and fire protection systems including portable fire extinguishers. As with healthcare occupancies, the compartmentation of the building and necessary provisions to ensure patient movement and evacuation are essential elements of ensuring occupant life safety. A thorough inspection of a facility falling under this occupancy class should also include determination of any hazardous areas, such as laboratories, as well as the life safety provisions that have been implemented to address corresponding life safety issues.

Day-Care Occupancy

NFPA 101 defines a **day-care occupancy** as an occupancy in which four or more clients receive care, maintenance, and supervision, by other than their relatives or legal guardians, for less than 24 hours per day (**FIGURE 2-9**). Examples of day-care occupancies include adult day-care occupancies, child day-care occupancies, day-care homes, kindergarten classes that are conducted within a day-care occupancy, and nursery schools. Day-care occupancies are not subject to the requirements governing educational occupancies. Day-care facilities within houses of worship are exempt during services.

The three subclasses of day-care occupancies are:

- Day-care occupancies
- Group day-care homes, generally within a dwelling unit; providing care to no fewer than 7 and no more than 12 clients at one time.
- Family day-care homes, a family day-care home in which more than 3 but fewer than 7 clients receive care.

FIGURE 2-9 An example of a day-care occupancy.

The three subclasses within the day-care occupancy category are distinguished based on the number of clients serviced. Day-care occupancies service the most clients, followed by group day-care homes, and family day-care homes. As a first responder inspector, it is important to recognize that the requirements for each of these occupancy subclasses may differ within the applicable code(s) within your jurisdiction. Clients related to the caregiver do not count towards the total census.

Considerations when inspecting a day-care occupancy include the occupant load, means of egress components: capacity of means of egress, the number of exits, the arrangement of means of egress, the travel distance to exits, discharge from exits, proper illumination of the means of egress, emergency lighting, the clear marking of the means of egress, and special means of egress features. Additional considerations when inspecting a day-care occupancy include the protection of vertical openings, protection from hazards, interior finishes, fire detection systems, alarm and communication systems, extinguishment requirements, special provisions, building services, and operating features.

Given the nature of these occupancies, especially in regards to the potential of sleeping occupants that require guidance for evacuation, it is important to ensure that the facility has appropriate separations and smoke detection, as required by code.

Educational Occupancy

Educational occupancies are defined by NFPA 101 as buildings used for educational purposes through the 12th grade by six or more persons that are occupied at least 4 hours per day or 12 hours per week (**FIGURE 2-10**). Examples of educational occupancies include academies, kindergartens, and schools. Schools

FIGURE 2-10 An example of an educational occupancy.

for levels beyond the 12th grade are not classified as educational occupancies. These facilities must comply with the requirements for assembly, business, or other appropriate occupancy. Day-care facilities are not classified as educational occupancies.

When the occupancy threshold of 50 or more persons is met, an educational occupancy may also be classified as an assembly occupancy. Educational facilities may contain assembly areas, such as auditoriums, cafeterias, and gymnasiums that typically present a low fire hazard but present a high concentration of occupants.

It is important to develop an understanding of the activities conducted within the building, the number of occupants, and the age of occupants as these factors may determine the occupancy classification. Educational activities can vary from the low fire hazards posed by the typical lecture classroom to the moderate or high fire hazards that a science laboratory presents. It should be noted that under some codes, the requirements for educational occupancies will be different for new buildings than for existing buildings. An example of this might be the requirements with respect to interior finishes.

Business Occupancies

According to NFPA 101, a **business occupancy** is as an occupancy used for the transaction of business other than mercantile. Businesses included under this occupancy classification are general offices, doctor's offices, government offices, city halls, municipal office buildings, courthouses, outpatient medical clinics where patients are ambulatory, college and university classroom buildings with less than 50 occupants, air traffic control towers, and instructional laboratories. Business occupancies typically have a large number of occupants during the business day and limited occupants during non-business hours (**FIGURE 2-11**).

Business occupancies generally have a lower occupant density than mercantile occupancies, and the occupants are usually more familiar with their surroundings. However, confusing and indirect egress paths are often developed due to office layouts and the arrangement of tenant spaces. When conducting an inspection of a business occupancy, you should consider means of egress, protection of openings, and hazardous areas. Computer rooms and the protection of the organization's records are business concerns that you will frequently encounter when inspecting a business occupancy. Building services such as waste disposal and the utilization of shafts and chases can contribute to fire protection problems. As always, you should inspect all fire protection systems for proper maintenance, testing, and operation. This would include alarm systems, sprinkler systems, and portable fire extinguishers.

Industrial Occupancies

An **industrial occupancy** is defined in NFPA 101 as an occupancy in which products are manufactured or in which processing, assembling, mixing, packaging, finishing, decorating, or repair operations are conducted (**FIGURE 2-12**). Industrial occupancies include, but are not limited to: chemical plants, factories, food processing plants, furniture manufacturers, hangars (for service and maintenance), laboratories involving hazardous chemicals, laundry and dry cleaning plants, metal-working plants, plastics manufacture and molding plants, power plants, refineries, semiconductor manufacturing plants, sawmills, pumping stations, telephone exchanges, and woodworking plants.

FIGURE 2-11 An example of a business occupancy.

FIGURE 2-12 An example of an industrial occupancy.

Industrial occupancies expose occupants to a wide range of processes and materials of varying hazard. Special-purpose industrial occupancies, which are characterized by large installations of equipment that dominate the space, such as a power plant, are addressed separately from general-purpose industrial facilities, which have higher densities of human occupancy. Industrial occupancy buildings, along with storage occupancy buildings, are more likely than any other occupancy to have contents with a wide range of hazards, often including an array of hazardous materials that can present fire and life safety hazards for building occupants and emergency response personnel.

Each building or separated portion should be inspected in accordance with the requirements of its primary use. The complexity of inspections of industrial occupancies can prove time consuming. Prior to conducting an inspection of an industrial occupancy, you should first determine and become familiar with all pertinent requirements under your jurisdiction's codes or regulations. Considerations during the inspection should include occupant load, means of egress, protection of openings, hazardous materials use and storage, inside storage, outside storage, housekeeping, and maintenance.

When inspecting these facilities, you should verify the maintenance and operational status of all fire protection systems including fire alarm systems, portable fire extinguishers, water supplies, fire pumps, standpipe systems, sprinkler systems, and special extinguishing systems.

Mercantile Occupancy

A **mercantile occupancy** is defined under NFPA 101 as an occupancy used for the display and sale of merchandise. Included under this occupancy classification are shopping centers, supermarkets, drug stores, department stores, auction rooms, restaurants with fewer than 50 persons, and any occupancy or portion thereof that is used for the display and sale of merchandise (**FIGURE 2-13**).

FIGURE 2-13 An example of a mercantile occupancy.

Mercantile occupancies, as in the case of assembly occupancies, are characterized by large numbers of people who gather in a space that is relatively unfamiliar to them. In addition, mercantile occupancies often contain sizable quantities of combustible contents and use circuitous egress paths that are deliberately arranged to force occupants to travel around displays of merchandise that is available for sale. Bulk merchandising retail buildings, which characteristically consist of a warehouse-type building occupied for sales purposes, are a subclass of mercantile occupancy with a greater potential for hazards than more traditional mercantile operations.

Mercantile occupancies vary in size and in terms of fire load. They encompass many different types of materials and operations. The separation between mercantile occupancies and other occupancies can be an issue that you must address. An example of this would be the shopping mall which contains stores, restaurants, and assembly areas. Covered malls and bulk merchandising retail centers present additional issues, such as fire spread, smoke travel, and challenges of evacuating building occupants, which must be considered during a fire inspection.

PRO TIPS

Although storage occupancies have low occupant loads, they may have very high fuel loads.

When inspecting a mercantile occupancy, you should consider occupant load, means of egress, interior finish, protection of openings, and protection from hazards. Fire protection systems are essential fire and life safety measures in mercantile occupancies. You should verify the operational status, maintenance, and testing of these systems.

Storage Occupancies

NFPA 101 defines a **storage occupancy** as an occupancy used primarily for the storage or sheltering of goods, merchandise, products, or vehicles. Storage occupancies are characterized by relatively low human occupancy in comparison to building size and by varied hazards associated with the materials stored. Storage occupancies include warehouses, freight terminals, parking garages, aircraft storage hangers, truck and marine terminals, bulk oil storage, cold storage, grain elevators, and barns (**FIGURE 2-14**).

PRO TIPS

Bulk merchandising retail buildings, which characteristically consist of a warehouse-type building occupied for sales purposes, are a subclass of mercantile occupancy rather than a storage occupancy.

Additionally, if occupant load is greater than normal for a storage occupancy, then the facility should be classified as an industrial occupancy. Examples of this are operations involving packaging, labeling, sorting, or special handling. Parking garages are classified as industrial occupancies if they contain an area in which repairs are performed.

When conducting an inspection of a storage occupancy, you should consider building contents, protection of openings, indoor and outdoor storage practices, hazardous materials and processes, and housekeeping. The occupant load and means of egress must be reviewed including: exit access and locations, identification of exits, and exit discharge.

FIGURE 2-14 An example of a storage occupancy.

The hazards associated with the operation of material-handling equipment, such as industrial fork lifts, and the storage and changing of associated fuel should be addressed in the inspection. In inspecting the fire protection systems, you should examine the fire alarm system, portable fire extinguishers, fire pumps, standpipe system, and sprinkler system.

Assembly Occupancies

NFPA 101 defines **assembly occupancies** as buildings that used for a gathering of 50 or more persons for deliberation, worship, entertainment, eating, drinking, amusement, awaiting transportation, or similar uses; or buildings that are used as a special amusement building regardless of occupant load (**FIGURE 2-15**). Assembly occupancies are characterized by the presence or potential presence of crowds with the potential of panic in the case of a fire or other emergency. They are generally or occasionally open to the public, and the occupants, who are present voluntarily, are not ordinarily subject to discipline or control. Such buildings are ordinarily occupied by able-bodied persons and are not used for sleeping purposes.

Assembly occupancies span a wide range of uses, each of which necessitates different considerations (**TABLE 2-1**). There are numerous legal ways an assembly occupancy can be utilized. An example of this could be a church, normally utilized as a place of worship, at times being used for dinning or entertainment. Just as the world is constantly changing, so too are the specific uses of the various buildings within your jurisdiction. Therefore it is important to recognize that both the character of an assembly occupancy and its occupant load may change over time and may have done so since the last time the building was inspected.

FIGURE 2-15 An example of an assembly occupancy.

TABLE 2-1 Examples of Assembly Occupancies

- Armories
- Assembly halls
- Auditoriums
- Bowling lanes
- Club rooms
- College and university classrooms (50 persons and over)
- Conference rooms
- Courtrooms
- Dance halls
- Drinking establishments
- Exhibition halls
- Gymnasiums
- Libraries
- Mortuary chapels
- Motion picture theatres
- Museums
- Passenger stations and terminals of air, surface, underground, and marine public transportation facilities
- Places of religious worship
- Pool rooms
- Recreation piers
- Restaurants
- Skating rinks
- Special amusement buildings (regardless of occupant load)
- Theatres

The primary considerations when inspecting assembly occupancy include verifying the occupant load, means of egress, interior finish requirements, building services, smoking provisions, and fire protection systems. As a first responder inspector, verify the posted occupant load, the availability and maintenance of the building's exits, and the building's fire protection systems.

In situations involving unique occupancies, such as stages, projection rooms, exhibits and trade shows, or special amusement buildings, it is important during the inspection to review any special safeguards that have been implemented, such as automatic sprinklers and exhaust systems. This is especially true in assembly occupancies where open flames or pyrotechnics may be present.

It must be recognized that while the probability of fire is usually low with assembly occupancies, the potential for loss of life can be high. Adequate exit provisions are thus a critical requirement for any assembly occupancy. The fire that occurred on February 20, 2003, at The Station, a nightclub in West Warwick, Rhode Island, illustrates the importance of the adequacy of fire exits in assembly occupancies. One hundred individuals perished in this tragic fire.

PRO TIPS

Some places of assembly are multi-use to the extent that the occupant density varies greatly from event to event. In hotels, the meeting rooms may be set up for classroom use during the day then change to a reception room with a dinner banquet at night. You must assure that the safeguards are adequate for the maximum number of people that may be expected in any such space under any conditions.

Detention and Correctional Occupancies

A **detention and correctional occupancy** is defined in NFPA 101 as an occupancy used to house one or more persons under varied degrees of restraint or security where such occupants are mostly incapable of self-preservation because of security measures not under the occupant's control (**FIGURE 2-16**). Facilities that fall under the detention and correctional occupancy classification include:

- Jails
- Detention centers
- Correctional institutions
- Houses of correction
- Prerelease centers

FIGURE 2-16 An example of a detention and correctional occupancy.

- Reformatories
- Work camps
- Training schools
- Substance abuse centers

There are five categories under this occupancy that correspond to the degree of restraint of occupants within the facility. These categories are:

- Use Condition I—Free Egress—Occupants are permitted to move freely to the exterior.
- Use Condition II—Zoned Egress—Occupants are permitted to move from sleeping areas to other smoke compartments.
- Use Condition III—Zoned Impeded Egress—Occupants are permitted to move within any smoke compartment.
- Use Condition IV—Impeded Egress—Occupants are locked in their rooms or cells, but they can be remotely unlocked.
- Use Condition V—Contained—Occupants are locked in their rooms or cells with manually operated locks that can only be opened by facility staff members.

Detention and correctional facilities present unique and challenging fire and life safety issues. The maintenance of a predefined level of security is essential within these facilities, thus facility personnel may be reluctant to initiate any actions, including evacuation procedures that may compromise security. Facility design, operation, and maintenance are crucial to ensuring fire and life safety, as well as prisoner security.

Within detention and correctional facilities, uses other than residential housing are classified according to the appropriate occupancy classification. The facility may consist of a number of buildings, some of which should be appropriately classified under different occupancies. Examples of this would be an infirmary which would be classified as a healthcare occupancy, an auditorium or gymnasium which would fall under the assembly occupancy classification, and a business office, storage area, or industrial shop which would be categorized under the appropriate occupancy classification.

Detention and correctional occupancies present the unique challenges associated with sleeping occupants and the occupant's lack of or limited control with respect to egress. Your role as a first responder inspector in ensuring the life safety of facility occupants cannot be overstated. The ability of the facility staff to implement a "defend in place" strategy successfully is based upon the design features of the building including its compartmentation, the maintenance and operational readiness of fire protection and other building systems, the existence of a comprehensive emergency plan, and the preparedness of staff members to execute the emergency plan based on their training, drills, and exercises. Under a defend in place strategy, which is typically used in those situations where conducting a total building evaluation would be problematic, building occupants are moved to and "sheltered" in protected and safe areas of refuge within the building.

Elements that must be considered when inspecting a facility falling under this occupancy classification include: capacity, means of egress, interior finish, protection of openings, contents, and hazardous areas. In reviewing means of egress the inspector should consider horizontal exits, sliding doors, remote controlled release mechanisms, and exit discharge.

Your inspection must include a thorough review of fire protection systems, including sprinkler and standpipe systems, building subdivision, and building services. In addition to inspecting the building fire protection systems, you should review accompanying documentation of system maintenance and testing.

Special Structures and High-Rise Buildings

Special structures and high-rise buildings present interesting challenges. High-rise buildings are typically considered buildings having occupied stories 75 ft (2286 cm) or more above the lowest level of fire department access to the building. Special structures include open structures, piers, towers, windowless buildings, underground structures, vehicles and vessels, water-surrounded structures, membrane structures, and tents (**FIGURE 2-17**).

FIGURE 2-17 An example of a special structure.
© Gail Johnson/Fotolia.com

A typical occupancy in a special structure will fall under one of the occupancy classifications previously covered. The means of egress and fire protection requirements for that occupancy apply, as do the additional requirements for the special structure. In addition to the requirements mandated by the applicable occupancy class, you must apply requirements unique to each special type of structure. Guidance on the requirements for special structures can be found in NFPA 101 and other relevant standards. When inspecting a special structure, you should first determine the general occupancy classification and then determine the special requirements that apply. Occupancies in special structures, while often not easy to inspect, must be inspected in accordance with all pertinent requirements.

WRAP-UP

CHAPTER SUMMARY

- In the model building codes, the term occupancy refers to the intended use of a building. The occupancy classification of a building is determined by the current use of that building.
- A one- or two-family dwelling unit is a building that contains no more than two dwelling units with independent cooking and bathroom facilities.
- A lodging or rooming house is a building or portion thereof that does not qualify as a one- or two-family dwelling, that provides sleeping accommodations for a total of 16 or fewer people on a transient or permanent basis, without personal care services, with or without meals, but without separate cooking facilities for individual occupants.
- A hotel is a building or group of buildings under the same management in which there are sleeping accommodations for more than 16 people and is primarily used by transients for lodging with or without meals.
- An apartment building is a building or portion thereof containing three or more dwelling units with independent cooking and bathroom facilities.
- A residential board and care occupancy as a building or portion thereof that is used for lodging and boarding of four or more residents, not related by blood or marriage to the owners or operators, for the purpose of providing personal care services.
- A healthcare occupancy as an occupancy used for purposes of medical or other treatment or care of four or more persons on an inpatient basis where such occupants are mostly incapable of self-preservation due to age, physical or mental disability, or because of security measures not under the occupant's control. These occupants also include infants, convalescents, or infirmed aged persons.
- Assembly occupancies are buildings that used for a gathering of 50 or more persons for deliberation, worship, entertainment, eating, drinking, amusement, awaiting transportation, or similar uses; or buildings that are used as a special amusement building regardless of occupant load.
- A day-care occupancy is an occupancy in which four or more clients receive care, maintenance, and supervision, by other than their relatives or legal guardians, for less than 24 hours per day.
- Educational occupancies are buildings used for educational purposes through the twelfth grade by six or more persons for 4 or more hours per day or more than 12 hours a week.
- A business occupancy is as an occupancy used for the transaction of business other than mercantile.
- An industrial occupancy is an occupancy in which products are manufactured or in which processing, assembling, mixing, packaging, finishing, decorating, or repair operations are conducted.
- A mercantile occupancy is an occupancy used for the display and sale of merchandise.
- A storage occupancy as an occupancy used primarily for the storage or sheltering of goods, merchandise, products, vehicles, or animals.
- Assembly occupancies are buildings that used for a gathering of 50 or more persons for deliberation, worship, entertainment, eating, drinking, amusement, awaiting transportation, or similar uses; or buildings that are used as a special amusement building regardless of occupant load.
- A detention and correctional occupancy is an occupancy used to house one or more persons under varied degrees of restraint or security where such occupants are mostly incapable of self-preservation because of security measures not under the occupant's control.

KEY TERMS

Ambulatory healthcare occupancy A building or portion thereof used to provide services or treatment simultaneously to four or more patients that, on an outpatient basis. (NFPA 101, *Life Safety Code*)

Apartment building is a building or portion thereof containing three or more dwelling units with independent cooking and bathroom facilities. (NFPA 101, *Life Safety Code*)

Assembly occupancies Buildings (1) used for a gathering of 50 or more persons for deliberation, worship, entertainment, eating, drinking, amusement, awaiting transportation, or similar uses; or (2) used as a special amusement building regardless of occupant load. (NFPA 101, *Life Safety Code*)

Business occupancy An occupancy used for the transaction of business other than mercantile. (NFPA 101, *Life Safety Code*)

Day-care occupancy An occupancy in which four or more clients receive care, maintenance, and supervision, by other than their relatives or legal guardians, for less than 24 hours per day. (NFPA 101, *Life Safety Code*)

Detention and correctional occupancy An occupancy used to one or more persons under varied degrees of restraint or security where such occupants are mostly incapable of self-preservation because of security measures not under the occupant's control. (NFPA 101, *Life Safety Code*)

Dormitory A building or space in a building in which group sleeping accommodations are provided for more than 16 persons who are not members of the same family in one room, or a series of closely associated rooms, under joint occupancy and single management, with or without meals, but without individual cooking facilities. (NFPA 101, *Life Safety Code*)

Educational occupancies Buildings used for educational purposes through the twelfth grade by six or more persons for 4 or more hours per day or more than 12 hours a week. (NFPA 101, *Life Safety Code*)

healthcare occupancy An occupancy used for purposes of medical or other treatment or care of four or more persons where such occupants are mostly incapable of self-preservation due to age, physical or mental disability, or because of security measures not under the occupant's control. (NFPA 101, *Life Safety Code*)

Hotel A building or group of buildings under the same management in which there are sleeping accommodations for more than 16 people and is primarily used by transients for lodging with or without meals. (NFPA 101, *Life Safety Code*)

Industrial occupancy An occupancy in which products are manufactured or in which processing, assembling, mixing, packaging, finishing, decorating, or repair operations are conducted. (NFPA 101, *Life Safety Code*)

Mercantile occupancy An occupancy used for the display and sale of merchandise. (NFPA 101, *Life Safety Code*)

Lodging or rooming house Building or portion thereof that does not qualify as a one- or two-family dwelling, that provides sleeping accommodations for a total of 16 or fewer people on a transient or permanent basis, without personal care services, with or without meals, but without separate cooking facilities for individual occupants. (NFPA 101, *Life Safety Code*)

Occupancy The intended use of a building

One- or two-family dwelling A building that contains no more than two dwelling units, each occupied by members of a single family with no more than three outsiders, if any, accommodated in rented rooms. (NFPA 101, Life Safety Code)

Residential board and care occupancy A building or portion thereof that is used for lodging and boarding of four or more residents, not related by blood or marriage to the owners or operators, for the purpose of providing personal care services. (NFPA 101, *Life Safety Code*)

Storage occupancy An occupancy used primarily for the storage or sheltering of goods, merchandise, products, vehicles, or animals. (NFPA 101, *Life Safety Code*)

Separated occupancy A multiple occupancy where the occupancies are separated by fire resistance-rated assemblies. (NFPA 101, *Life Safety Code*)

You Are the First Responder Inspector

You prepare to make an inspection at an existing surf board shop in a strip mall. You look in the file and locate the folder which classifies the occupancy as mercantile. When you enter the shop, you smell a strong chemical odor. You find that the business now has begun repairing surf boards in the back room in addition to sales. The operation involves using different flammable resins, paints, and solvents. Also, different power tools that produce large amounts combustible dusts are in use. You make notes in preparation of correcting the occupancy classification.

1. **What single factor would change the classification of this occupancy?**
   - **A.** Use of flammable liquids
   - **B.** Combustible dusts
   - **C.** Repair operations
   - **D.** All of the above
2. **What is the correct classification of this occupancy?**
   - **A.** Assembly
   - **B.** Business
   - **C.** Industrial
   - **D.** Storage
3. **What can the business owner do to keep the mercantile occupancy classification?**
   - **A.** The business location must be changed.
   - **B.** Eliminate the repair shop.
   - **C.** Move all operations except sales to another location.
   - **D.** Either B or C.
4. **If you select the wrong occupancy class, what is the consequence?**
   - **A.** The incorrect licensing fees will be charged.
   - **B.** An inadequate level of fire and life safety may be required.
   - **C.** Insurance coverage may be denied.
   - **D.** Prefire planning will be incorrect.

CHAPTER

3

Fire Growth

NFPA 1030 THAT INCLUDES CHAPTER 6: FIRST RESPONDER INSPECTOR (NFPA 1031)

First Responder Inspector

- 6.5.10

KNOWLEDGE OBJECTIVES

After studying this chapter, you will be able to:

1. Describe the ignition phase, growth phase, fully developed phase, and decay phase of a fire.
2. Describe the characteristics of a room-and-contents fire through each of the four phases of a fire.
3. Describe how interior finishes affect fire growth.
4. Describe how decorative materials may affect fire growth.
5. Describe how furnishings may affect fire growth.
6. Describe how building construction elements may affect fire growth.
7. Describe how to obtain the flame spread and smoke development ratings for the contents of an occupancy.

SKILLS OBJECTIVES

There are no skills objectives for this chapter.

You Are the First Responder Inspector

You are inspecting a nightclub that has been redecorated. In the VIP lounge, you are surprised to see wall-to-wall and floor-to-ceiling carpeting in a windowless room. In your experience, the owners will be more receptive to your suggestions if you explain why a change needs to be made. You clear your throat and prepare to describe in simple and understandable terms how fire growth is affected by interior finishes.

1. In your own words, describe the concept of the fire tetrahedron.
2. Explain how heat travels and why.
3. Describe the how the interior finishes of this room will contribute to fire growth.

Introduction

The first responder inspector (FRI) is expected to ensure code compliance and identify situations where the fire triangle or the fire tetrahedron can become operational and start a fire. Once that fire occurs, the materials of construction and occupancy become potential enhancement to fire growth. As the first responder inspector, you are expected to be able to analyze situations, identify potential adverse situations, and act accordingly.

The Chemistry of Fire

To be effective as a first responder inspector, you need to understand the conditions needed for a fire to ignite and grow. Understanding these conditions will enable you to identify potential fire hazards in the structures you inspect.

What Is Fire?

As defined by NFPA 921, **fire** is a rapid oxidation process, which is a chemical reaction resulting in the evolution of light and heat in varying intensities. We all know it when we see it, but we really do not know much about how fire works. A unique phenomenon, fire comes in different colors, ranging from red to yellow to blue. Flames can be steady, they can flicker, or they can just glow. We can see flames, but we can also see through them. Fire is neither solid nor liquid. Wood is a solid, gasoline is a liquid, and propane is a gas—but they all burn.

States of Matter

Matter is a term used to describe any physical substance that has mass and volume. Matter is made up of atoms and molecules. Matter exists in three states: solid, liquid, and gas (**FIGURE 3-1**). An understanding of these states of matter is helpful in understanding fire behavior.

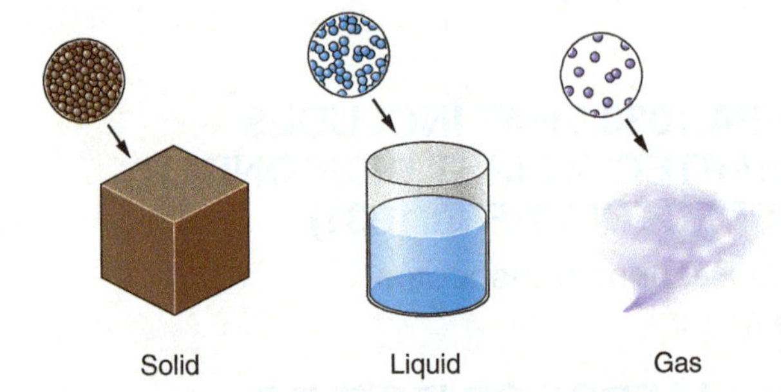

FIGURE 3-1 Fuels can be classified as solids, liquids, or gases.

We all know what makes an object solid: it has a definite shape. Most uncontrolled fires are stoked by solid fuels. In structure fires, the building and most of the contents are **solids**. A solid does not flow under stress. Instead, it has a definite capacity for resisting forces and under ordinary conditions retains a definite size and shape. One characteristic of a solid is its ability to expand when heated and to contract when cooled. In addition, a solid object may change to a liquid state or a gaseous state when heated. Cold makes most solids more brittle, whereas heat makes them more flexible. Because a solid is rigid, only a limited number of molecules are on its surface; the majority of molecules are cushioned or insulated by the outer surface of the solid.

A **liquid** will assume the shape of the container in which it is placed. Most liquids contract when cooled and expand when heated. In addition, most will turn into gases when sufficiently heated. Liquids, for all practical purposes, do not compress. This characteristic allows firefighters to pump water for long distances through pipelines or hoses. A liquid has no independent shape, but it does have a definite volume. The liquid with which firefighters are most concerned is water.

A **gas** is a type of liquid that has neither independent shape nor independent volume, but rather tends to expand indefinitely. The gas we most commonly encounter is air, the mix of invisible odorless, tasteless gases that surrounds the earth. The mixture of gases in air maintain a constant composition—21 percent oxygen, 78 percent nitrogen, and 1 percent other gases such as carbon dioxide. Oxygen is required for us to live, but it is also required for fires to burn. We will explore the reaction of fuels with oxygen as we look at the chemistry of burning. Some fuels exist in the form of a gas.

Fuels

A **fuel** is any material that stores potential energy. Think of the vast amount of heat that is released during a large fire. The energy released in the form of heat and light has been stored in the fuel before it is burned. The release of the energy in a gallon of gasoline, for example, can move a car miles down the road (**FIGURE 3-2**).

Types of Energy

Energy exists in the following forms: chemical, mechanical, electrical, light, or nuclear. Regardless of the form in which the energy is stored, it can be changed from one form to another. For example, electrical energy can be converted to heat or to light. Mechanical energy can be converted to electrical energy through a generator.

Chemical Energy

Chemical energy is the energy created by a chemical reaction. Some chemical reactions produce heat (**exothermic**); others absorb heat (**endothermic**). The combustion process is an exothermic reaction, because it releases heat energy. Most chemical reactions occur because bonds are established between two substances, or bonds are broken as two substances are chemically separated. Heat is also produced whenever oxygen combines with a combustible material. If the reaction occurs slowly in a well-ventilated area, the heat is released harmlessly into the air. If the reaction occurs very rapidly or within an enclosed space, the mixture can be heated to its **ignition temperature** and can begin to burn. A bundle of rags soaked with linseed oil will begin to burn spontaneously, for example, because of the heat produced by oxidation that occurs within the mass of rags. The combustion potential of gasoline is one example of chemical energy.

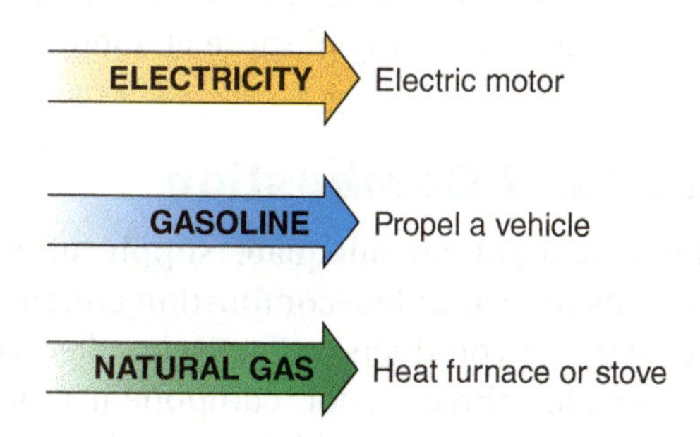

FIGURE 3-2 Energy being converted to work.

Mechanical Energy

Mechanical energy is converted to heat when materials rub against each other and create friction. For example, a fan belt rubbing against a seized pulley produces heat. Water falling across a dam is one example of mechanical energy based on position: the water releases energy as it moves to a lower area.

PRO TIPS

A liquid expands when heated and contracts when cooled. Water expands when heated and also when it freezes. This is why frozen pipes and fittings burst when they freeze.

Electrical Energy

Electrical energy is converted to heat energy in several different ways. For example, electricity produces heat when it flows through a wire or any other conductive material. The greater the flow of electricity and the greater the resistance of the material, the greater the amount of heat produced. Examples of electrical energy that can produce enough heat to start a fire include heating elements, overloaded wires, electrical arcs, and lightning. Electrical energy is carried through the electrical wires inside homes and is stored in batteries that convert chemical energy to electrical energy.

Light Energy

Light energy is caused by electromagnetic waves packaged in discrete bundles called photons. This energy travels as thermal radiation, a form of heat. When light energy is hot enough, it can sometimes be seen in the form of visible light. If it is of a frequency that we cannot see, the energy may be felt as heat but not seen as visible light. Candles, fires, light bulbs, and lasers are all forms of light energy. Another example of light energy is the radiant energy we receive from the sun.

Nuclear Energy

Nuclear energy is created by splitting the nucleus of an atom into two smaller nuclei (nuclear fission) or by combining two small nuclei into one large nucleus (fusion). Nuclear reactions release large amounts of energy in the form of heat. These reactions can be controlled, as in a nuclear power plant, or uncontrolled, as in an atomic bomb explosion. In a nuclear power plant, the nuclear reaction releases carefully controlled amounts of heat, which then convert water to steam which powers a steam turbine generator. Both explosions and controlled reactions release radioactive material, which can cause injury or death. Nuclear energy is stored in radioactive materials and converted to electricity by nuclear power-generating stations.

Conservation of Energy

The law of the conservation of energy states that energy cannot be created or destroyed; however, it can be converted from one form to another. Think of an automobile: chemical energy in the gasoline is converted to mechanical energy when the car moves down the road. If you apply the brakes to stop the car, the mechanical energy is converted to heat energy by the friction of the brakes, and the car slows. Similarly, in a house fire, the stored chemical energy in the wood of the house is converted into heat and light energy during the fire.

PRO TIPS

A photo cell in a standard solar backyard light collects energy from sunlight and converts it into electricity to power the light.

Conditions Needed for Fire

To understand the behavior of fire, you need to consider the three basic elements needed for a fire to occur: fuel, oxygen, and heat. First, a combustible fuel must be present. Second, oxygen must be available in sufficient quantities. Third, a source of ignition, often heat, must be present. If we graphically place these three together, the result is a **fire triangle** (**FIGURE 3-3**).

Once a fire has ignited, four factors are required to maintain a self-sustaining fire. These are fuel, oxygen, heat, and chain reaction. That is, the fuel, oxygen, and heat continue to interact so that a chain reaction keeps the fire going.

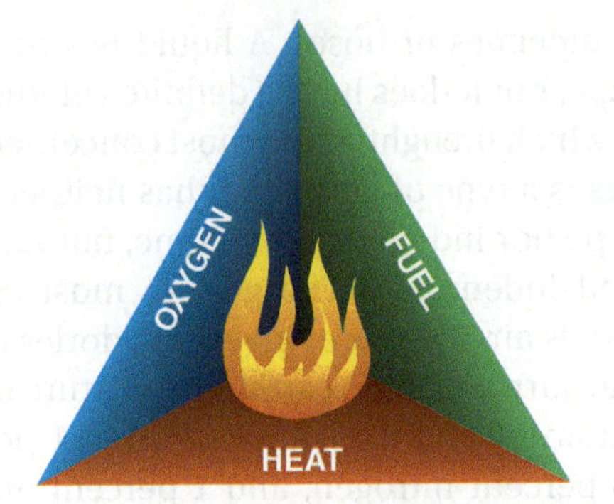

FIGURE 3-3 The fire triangle consists of fuel, oxygen, and heat.

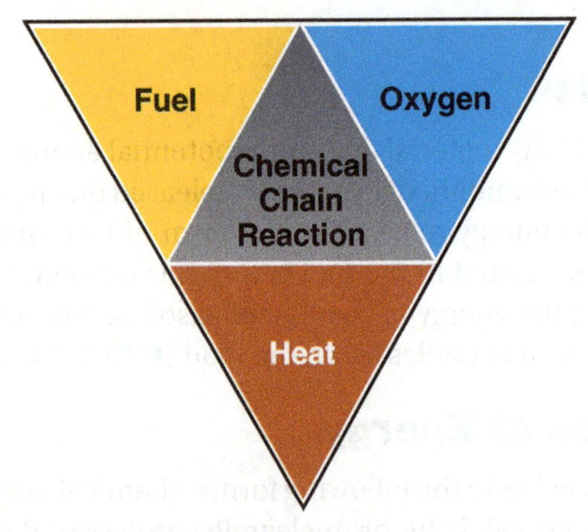

FIGURE 3-4 The tetrahedron model of fire (the fire tetrahedron) includes chemical chain reactions as an essential part of the combustion process.

Another way of visualizing all four conditions needed for a fire is the **fire tetrahedron** (**FIGURE 3-4**). A tetrahedron is a four-sided, three-dimensional figure. Each side of the fire tetrahedron represents one of the four conditions needed for a fire to occur.

Either of these visualizations will help you remember the four conditions that must be present for a fire to occur. The key point is that fuel, oxygen, and heat must be present for a fire to start and to continue burning. If you remove any of these elements, the fire will go out.

Products of Combustion

Fires burn without an adequate supply of oxygen, which results in incomplete combustion and produces a variety of toxic byproducts, collectively called **smoke**. Smoke includes three major components: particles (which are solids), vapors (which are finely suspended liquids—that is, aerosols), and gases.

Smoke particles include unburned, partially burned, and completely burned substances. The unburned particles are lifted in the **thermal column** produced by the fire. The thermal column is the cylindrical area above a fire in which heated air and gases rise and travel upward. Some partially burned particles become part of the smoke because inadequate amount of oxygen is available to allow for their complete combustion. Completely burned particles are primarily ash. Some particles in smoke are small enough that they can get past the protective mechanisms of the respiratory system and enter the lungs, and most of these are very toxic to the body.

Smoke may also contain small droplets of liquids (**FIGURE 3-5**). The fog that is formed on a cool night consists of small water droplets suspended in the air. When water is applied to a fire, small droplets may also be suspended in the smoke or haze that forms. Similarly, when oil-based compounds burn, they produce small oil-based droplets that become part of the smoke. Oil-based or lipid compounds can cause great harm to a person when they are inhaled. In addition, some toxic droplets cause poisoning if absorbed through the skin.

Smoke contains a wide variety of gases. The composition of gases in smoke will vary greatly, depending on the amount of oxygen available to the fire at that instant. The composition of the substance being burned also influences the composition of the smoke. In other words, a fire fueled by wood will produce a different composition of gases than a fire fueled by petroleum-based fuels, including plastics.

Almost all of the gases produced by a fire are toxic to the body, including carbon monoxide, hydrogen cyanide, and phosgene. Carbon monoxide is deadly in small quantities and was used to kill people in gas chambers. Hydrogen cyanide was also used to kill convicted criminals in gas chambers. Phosgene gas was used in World War I as a poisonous gas to disable soldiers. Even carbon dioxide, which is an inert gas, can displace oxygen and cause **hypoxia**. Hypoxia is a state of inadequate oxygenation of the blood and tissue.

A discussion of the by-products of combustion would be incomplete without considering heat. Because smoke is the result of fire, it is hot. The temperature of smoke will vary depending on the conditions of the fire and the distance the smoke travels from the fire. Injuries from smoke may occur because of the inhalation of the particles, droplets, and gases that make up smoke. The inhalation of superheated gases in smoke may also cause injury and burns in the respiratory tract. Intense heat from smoke can also severely burn the skin.

Fire Spread

Fires grow and spread by three primary mechanisms: conduction, convection, and radiation.

Conduction

Conduction is the process of transferring heat through matter by movement of the kinetic energy from one particle to another (**FIGURE 3-6**). Conduction transfers energy directly from one molecule to another, much as a billiard ball transfers energy from one ball to the next. Objects vary in their ability to conduct energy. Metals generally have a greater ability to conduct heat than wood does, whereas a substance such as fiberglass (which is used for insulation) has almost no ability to conduct heat or fire.

Objects that are good conductors tend to absorb heat and conduct it throughout the object. For example, heat applied to a steel beam will be readily conducted along the beam. Because the heat spreads out

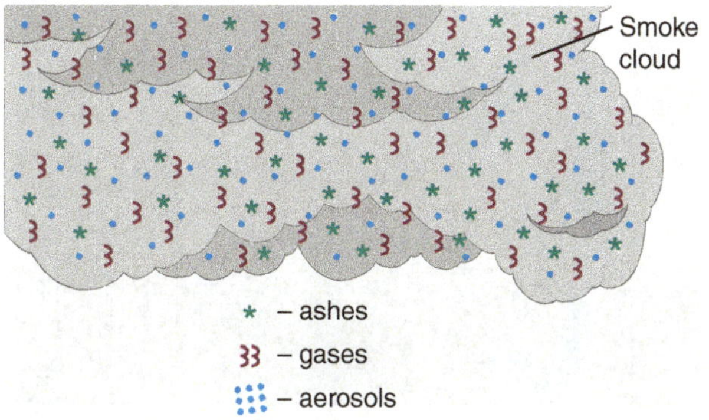

Smoke cloud

* – ashes

33 – gases

– aerosols

FIGURE 3-5 Smoke is composed of solids, aerosols (vapors), and gases.

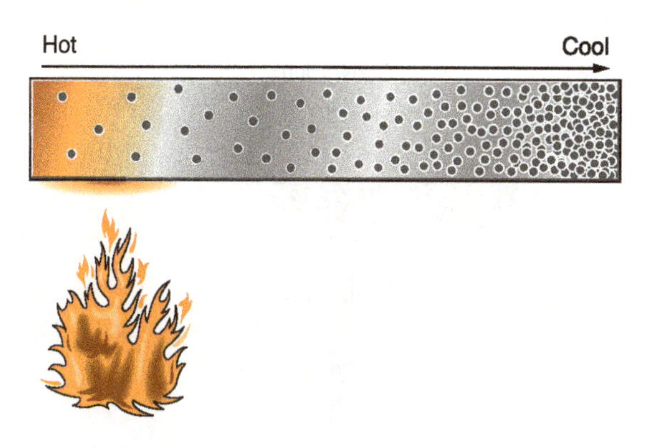

FIGURE 3-6 Conduction.

over the beam, the area of the steel beam to which the heat is being applied will not get as hot as it would if the beam were a poor conductor.

Applying heat to a poor conductor such as wood will result in the heat energy staying in the area of the wood to which the heat is applied. Because the wood has a fairly low ignition temperature, the wood will ignite. Heat applied to the wood acts on a small part of the wood; the heat is not conducted to other parts of the wood. By comparison, the same amount of heat energy applied to the steel beam will result in less intense heating at the point where the heat is applied because much of the heat dissipates to other parts of the steel beam.

This behavior has serious consequences for first responder inspectors: If two substances have the same ignition temperature, it will be easier to ignite the substance that is a poor conductor. The most important fact to remember about conductivity and fire spread is that poor conductors may ignite more easily but, once ignited, do not spread fire through conduction. Materials that are good conductors are rarely the primary means of spreading a fire.

Convection

Convection is the circulatory movement that occurs in a gas or fluid with areas of differing temperatures owing to the variation of the density and the action of gravity (**FIGURE 3-7**). To understand the movement of gases in a fire, consider a container of water being heated on a stove. If you put a drop of food coloring in one part of the water and start to heat the water, you can readily see the circulation of the water within the container. As the water heats up, it expands and becomes lighter than the surrounding water. This causes a column of water to rise to the top; the cooler water then falls to the bottom. Thus a cycle of warmer water rises to the top, pushing the cooler water to the bottom. The convection currents have the same effect of a small pump pushing the water around in a set pathway.

The convection currents in a fire involve primarily gases generated by the fire. The heat of the fire warms the gases and particles in the smoke. A large fire burning in the open can generate a **plume** of heated gases and smoke that rises high in the air. This convection stream can carry smoke and large bands of burning fuel for several blocks before the gases cool and fall back to the earth. If winds are present during a large fire, they may influence the direction in which the convection currents travel.

PRO TIPS

The vertical movement of the heated gases is commonly referred to as mushrooming due to the shape of the smoke. When the hot, burning gases travel horizontally along the ceiling, it is commonly referred to as rollover or flameover.

When a fire occurs in a building, the convection currents generated by the fire rise in the room and travel along the ceiling. These currents carry superheated gases, which may ultimately heat flammable materials enough to ignite them (**FIGURE 3-8**). If the fire room has openings, convection may then carry the fire outside the room of origin and to other parts of the building.

Radiation

Radiation is the transfer of heat through the emission of energy in the form of invisible waves (**FIGURE 3-9**). The sun radiates energy to the earth over the vast miles of outer space; the electromagnetic radiation readily

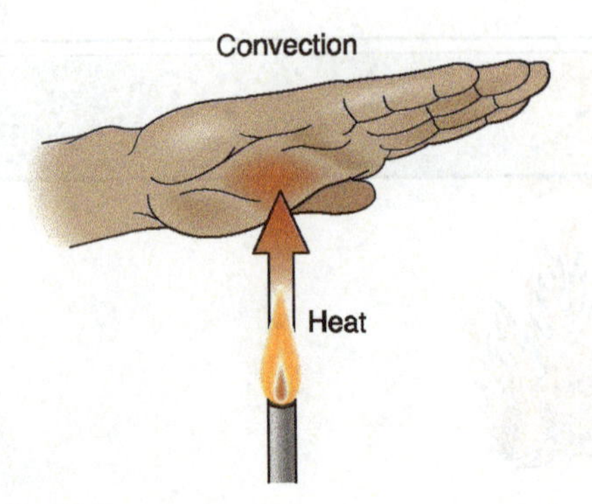

FIGURE 3-7 Convection.

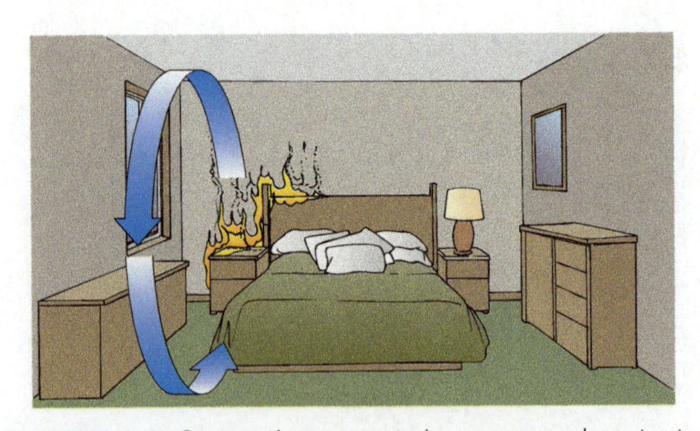

FIGURE 3-8 Convection currents in a room-and-contents fire.

FIGURE 3-9 Radiation.

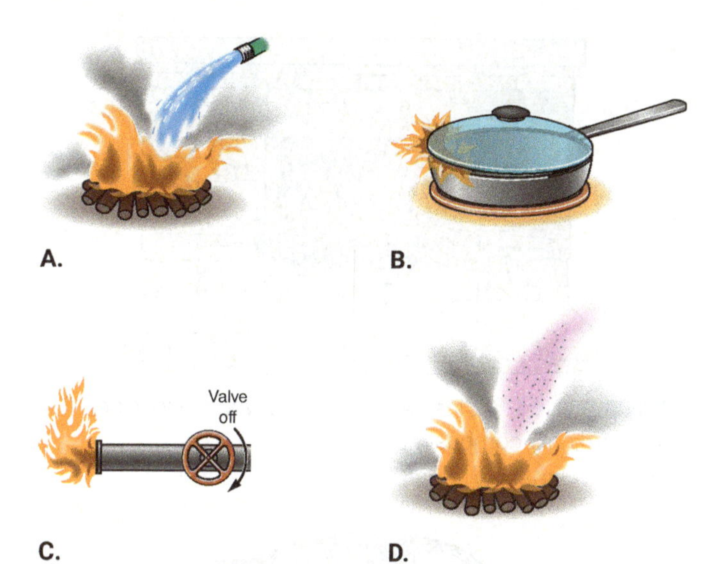

FIGURE 3-10 The four basic methods of fire extinguishment. **A.** Cool the burning material. **B.** Exclude oxygen from the fire. **C.** Remove fuel from the fire. **D.** Interrupt the chemical reaction with a flame inhibitor.

travels through the vacuum of space. When this energy is absorbed, it become converted to heat—that is, the heat you feel as the sun touches your body on a warm day. Of course, the direction in which the radiation travels can be changed or redirected, as when a sheet of shiny aluminum foil reflects the sun's rays and bounces the energy in another direction.

Thermal radiation from a fire travels in all directions. The effect of thermal radiation, however, is not seen or felt until the radiation strikes an object and heats the surface of the object. Thermal radiation is a significant factor in the growth of a camp-fire from a small flicker of flame to a fire hot enough to ignite large logs. The growth of a fire in a wastebasket to a room-and-contents fire is due in part to the effect of thermal radiation. A building that is fully involved in fire radiates a tremendous amount of energy in all directions. Indeed, the radiant heat from a large building fire can travel several hundred feet to ignite an unattached building.

Methods of Extinguishment

Although there are many variations on the methods used to extinguish fires, they boil down to four main methods: cooling the burning material, excluding oxygen from the fire, removing fuel from the fire, and interrupting the chemical reaction with a flame inhibitor (**FIGURE 3-10**). There are many variations on the way that these methods can be implemented, and sometimes a combination of these methods is used to achieve suppression of fires.

The method most commonly used to extinguish fires is to cool the burning material. Pouring water onto a fire is one of the most common ways that firefighters decrease fire temperature.

A second method of extinguishing a fire is to exclude oxygen from the fire. One example is placing the lid on an unvented charcoal grill. Likewise, applying foam to a petroleum fire excludes oxygen from the burning fuel.

Removing fuel from a fire will extinguish the fire. For example, shutting off the supply of natural gas to a fire being fueled by this gas will extinguish the fire. In wildland fires, a firebreak cut around a fire puts further fuel out of the fire's reach.

The fourth method of extinguishing a fire is to interrupt the chemical reaction with a flame inhibitor. Halon-type fire extinguishers, often used to protect computer systems, work in this way. These agents can be applied with portable extinguishers or through a fixed system designed to flood an enclosed space.

Classes of Fire

Fires are generally categorized into one of five classes: Class A, Class B, Class C, Class D, and Class K. Any structure may be at risk for any of the classes of fire depending upon the type of construction, occupancy, and materials or process in use in the structure.

Class A Fires

Class A fires involve solid combustible materials such as wood, paper, and cloth (**FIGURE 3-11**). Natural vegetation such as the grass that burns in brushfires is also considered to be part of this group of materials. The

FIGURE 3-11 A Class A fire involves wood, paper, or other ordinary combustibles.

FIGURE 3-12 A Class B fire involves flammable liquids such as gasoline.

method most commonly used to extinguish Class A fires is to cool the fuel with water to a temperature that is below the ignition temperature.

Class B Fires

Class B fires involve flammable or combustible liquids such as gasoline, kerosene, diesel fuel, and motor oil (**FIGURE 3-12**). Fires involving gases such as propane or natural gas are also classified as Class B fires. These fires can be extinguished by shutting off the supply of fuel or by using foam to exclude oxygen from the fuel.

Class C Fires

Class C fires involve energized electrical equipment (**FIGURE 3-13**). They are listed in a separate class because of the electrical hazard they present. Attacking a Class C fire with an extinguishing agent that conducts electricity can result in injury or death to the firefighter. Once the power is cut to a Class C fire, the fire can be treated as a Class A or Class B fire depending on the type of material that is burning.

Class D Fires

Class D fires involve combustible metals such as sodium, magnesium, and titanium (**FIGURE 3-14**). These

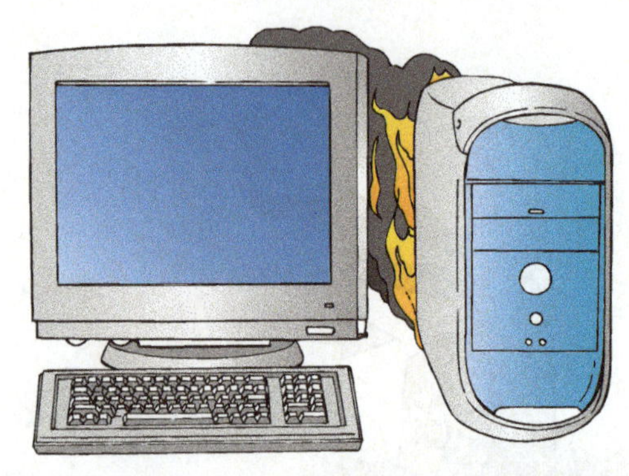

FIGURE 3-13 A Class C fire involves energized electrical equipment.

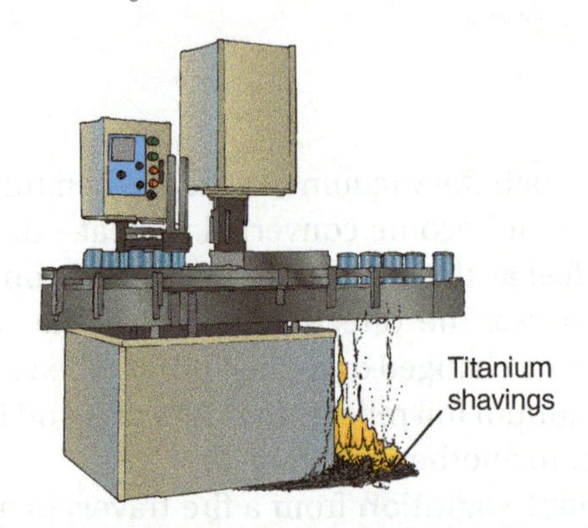

FIGURE 3-14 A Class D fire involves metals such as sodium or magnesium.

FIGURE 3-15 A Class K fire involves a deep-fat fryer in a kitchen.

fires are assigned to a special class because the application of water to fires involving these metals will result in violent explosions. These fires must be attacked with special agents to prevent explosions and to achieve fire suppression.

Class K Fires

Class K fires involve combustible cooking oils and fats in kitchens (**FIGURE 3-15**). Heating vegetable or animal fats or oils in appliances such as deep-fat fryers can result in serious fires that are difficult to fight with ordinary fire extinguishers. Special Class K extinguishers are available to handle this type of fire.

A fire may fit into more than one class. For example, a fire involving a wood building could also involve petroleum products. Likewise, a fire involving energized electrical circuits—a Class C fire—might also involve Class A or Class B materials. If a fire involves live electrical sources, it should be treated as a Class C fire until the source of electricity has been disconnected.

Characteristics of Solid-Fuel Fires

This section uses wood as an example of a solid fuel because it is the most common solid fuel and because it shares many characteristics with other solid fuels.

Solid fuels have a definite form and a defined shape. Wood conducts little heat, so the heat acts only on its surface. As a result, the fuel reaches its ignition temperature quickly and ignites. A thin piece of wood burns quickly because of the large surface area exposed to the heat and air. The thinner the wood is cut, the faster it ignites and burns. Thin wood is exposed to heat from both sides, which explains why it is easier to ignite a matchstick than a large log.

Solid fuels do not actually burn in the solid state. Instead, a solid fuel must be instead heated or pyrolyzed to decompose it into a vapor before it will burn. Many solid fuels change directly from a solid to a gas; others change from the solid state to a liquid before they are vaporized.

Because solid fuels do not generally conduct heat, a heat source applied to the surface of the fuel will cause decomposition of only the surface of the material. The ignition of any solid fuel is related to the temperature of the surface of the material as well as the vapors released through pyrolysis. The temperature of the interior of the material, however, does not affect the speed with which the material can be ignited. If the fuel conducted heat to other parts of the fuel, it would take more heat to raise the temperature of the material to the ignition temperature.

Wood does not have a fixed ignition temperature. Rather, the temperature at which wood will ignite depends on the presence of enough flammable vapors to produce the burning reaction. The ignition temperature varies according to the manner and rate at which the heat is applied.

Consider paper as a fuel. Paper is made from wood. The basic ingredient of paper is cellulose, the same ingredient that is found in wood. The ignition temperature of paper is 425°F. Paper ignites because its surface temperature increases rapidly, which allows for rapid burning and a high but brief heat release rate.

SAFETY TIPS

Do not attempt to extinguish any fire that is energized with electricity with any extinguishing agent not suitable for Class C fires.

PRO TIPS

The thermal conductivity of wood varies with the species and the temperature. The thermal conductivity values for engineered wood is higher than for conventional wood.

PRO TIPS

The configuration of the fuel determines how rapidly the fuel will yield its stored energy. An one pound log contains approximately 8000 BTUs. One pound of pine needles also contains approximately 8000 BTUs. Which fuel will release its heat fastest? This is called the rate of heat release. The material with the greatest surface area is heated more easily reaches its ignition temperature faster and is converted to gas faster. At the same time, oxygen in the air can readily combine with the flammable gas to create a combustible mixture and undergo combustion. The fire will be brief but severe.

Solid-Fuel Fire Development

Solid-fuel fires progress through four phases: the ignition phase, the growth phase, the fully developed phase, and the decay phase. To illustrate the behavior of a fire through these phases, we will use a wooden dollhouse as an example. We will make three assumptions:

- The dollhouse is constructed on a sandy surface that will not burn.
- There are no exposures to which the fire can spread.
- There is no wind to influence the behavior of the fire.

Ignition Phase

The **ignition phase**, in which the fire is limited to its point of origin, begins as a lighted match is placed next to a crumpled piece of paper. The heat from the match ignites the paper, which sends a small plume of fire upward (**FIGURE 3-16**). The heat generated from the paper sets up a small convection current, and the

FIGURE 3-16 The ignition phase of a fire.

FIGURE 3-17 The growth phase of a fire.

flame produces a small amount of radiated energy. The combination of convection and radiation serve to heat the fuel around the paper. Because the convection acts more in an upward direction, the kindling above the paper will be heated to the ignition temperature first. It is small, so its surface can be easily heated enough to release flammable vapors from the wood.

PRO TIPS

In a house fire, the growth phase is the point in the fire development that a smoke detector would operate to alert the occupants to evacuate. If sprinklers are installed, the sprinkler would operate, discharging water on the fire, either extinguishing the fire or controlling it until the fire department arrives.

Growth Phase

The **growth phase**, when the fire spreads to nearby fuel, occurs as the kindling starts to burn, increasing the convection of hot gases upward (**FIGURE 3-17**). The hot gases and the flame both act to raise the temperature of the wood located above the kindling. Energy generated by the growing fire starts to radiate in all directions. The convection of hot gases and the direct contact with the flame cause major growth to occur in an upward direction. The energy radiated by the flame will also cause some growth of the fire in a lateral direction. The major growth will be in an upward direction, however, because the radiation and convection both act in an upward direction, whereas growth beside the flame is primarily a function of radiation alone. The net result of the growth phase is that the fire grows from a tiny flame to a fire that involves the heavier wood.

FIGURE 3-18 The fully developed phase of a fire.

Fully Developed Phase

The **fully developed phase** produces the maximum rate of burning (**FIGURE 3-18**). All available fuel has ignited and heat is being produced at the maximum rate. At the fully developed phase, thermal radiation extends in all directions around the fire. The fire can spread downward by radiation and by flaming material that falls on unburned fuel. The fire can spread upward through the thermal column of hot gases and flame above the fire. In fact, a large free-burning fire can carry a thermal column several hundred feet into the air.

A fully developed fire burning outside is limited only by the amount of fuel available. By comparison, a fully developed fire in a building may be limited by the amount of oxygen that is available to it. The fully developed phase lasts as long as a large supply of fuel is available. The temperatures of the fire will vary from one part of the flame to another.

Decay Phase

The final phase of a fire is the **decay phase**, the period when the fire is running out of fuel (**FIGURE 3-19**). During the decay phase, the rate of burning slows down

FIGURE 3-19 The decay phase of a fire.

because less fuel is available. The rate of thermal radiation decreases. The amount of hot gases rising above the fire also decreases because less heat is available to push the gases aloft and smaller amounts of gases are being produced by the decreasing volume of fire. Eventually, the flames become smoldering embers and the fire goes out because of a lack of fuel.

Key Principles of Solid-Fuel Fire Development

Several important principles of fire behavior are illustrated by the dollhouse fire:

- Hot gases and flame are lighter and tend to rise.
- Convection is the primary factor in spreading the fire upward.
- Downward spread of the fire occurs primarily from radiation and falling chunks of flaming material.
- If there is not more fuel above or beside the initial flame that can be ignited by convection or radiated heat, the fire will go out.
- Variations in the direction of upward fire spread will occur if air currents deflect the flame.
- The total material burned reflects the intensity of the heat and the duration of the exposure to the heat.
- An adequate supply of oxygen must be available to fuel a free-burning fire, although some parts of the flame may have a limited supply of oxygen.

The dollhouse example focused on a relatively simple form of fire. Fire behavior becomes more complex, of course, when fires occur within an actual building.

Characteristics of a Room-and-Contents Fire

A fire in a building is not just a fire brought inside. The construction of the building and the contents of the room have a major impact on the behavior of the fire.

Room Contents

Fifty years ago, rooms contained many products made from natural fibers and wood. Today, synthetic products are all around us. This change in room contents has changed the behavior of fires. The synthetic products that are so prevalent today—especially plastics—are usually made from petroleum products. When heated, most plastics will pyrolyze to volatile products. These by-products of the combustion process are usually not only flammable, but also toxic. Some will melt and drip. If these droplets are on fire, they can contribute to the spread of the fire. Burning plastics generate dense smoke that is rich in flammable vapors, which makes fire suppression both more difficult and more dangerous.

Walls and ceilings are often painted. Newer paints are generally emulsions of latex, acrylics, or polyvinyl. When these paints are applied, they form a plastic-like coating. This coating has the characteristics of the plastic from which it was formed, and burns readily. Similarly, varnishes and lacquers are very combustible. Most paints and coatings will add to the fire present in a building, and their presence may aid in the spread of the fire from one room to another.

Carpets made today have low melting points and ignite readily. The backing of carpets is often polypropylene; polyurethane foam is commonly used as well. Carpet is readily ignitable by radiant heat, even when it is some distance away from a fire.

Furniture manufactured today has improved resistance to heat from glowing sources such as cigarettes, but it has almost no resistance to ignition from flaming sources. Such furniture can be completely involved in fire in 3 to 5 minutes, and it may be reduced to a burning frame in just 10 minutes.

In summary, the increased use of plastics in furniture, bedding, paints, and carpets sets the stage for complex fires. To see how such a fire might evolve, we will consider a fire in an ordinary bedroom that has painted walls and wall-to-wall carpeting. The room is furnished with a bed, two nightstands, a lamp, a vanity, and a wastebasket.

Ignition Phase

The ignition phase begins when a fallen candle ignites the contents crumpled in the plastic wastebasket sitting next to the vanity (**FIGURE 3-20**). The flame begins small, with a localized flame. This phase typically produces an open flame. The percentage of oxygen present

PRO TIPS

Outside influences can change the course of a fire. Fuel value, fuel configuration, oxygen levels, drafts and winds, and humidity are just a few variables that can influence fire growth. You must be familiar with these variables and never discount them during an inspection.

FIGURE 3-20 The ignition phase of a typical room-and-contents fire.

FIGURE 3-21 The growth phase of a typical room-and-contents fire.

at this point is 21 percent—the same as the percentage in the room air.

As more of the contents in the wastebasket are ignited, a plume of hot gases rises from the wastebasket. Combustible materials in the path of the flame begin to ignite, which increases the extent and intensity of the flame. The convection of hot gases is the primary means of fire growth. Some energy is radiated to the area close to the flame, so that the plastic wastebasket starts to melt and may ignite. Oxygen is drawn in at the bottom of the flame above the wastebasket. At this point, the fire could probably be extinguished with a portable fire extinguisher.

Growth Phase

During the growth phase, additional fuel is drawn into the fire (**FIGURE 3-21**). As more fuel is ignited, the size of the fire increases. The plume of hot gases and flames creates a convection current that carries hot gases to the ceiling of the room. Next, flammable materials in the path of this plume ignite. Flames begin to spread upward and outward. Hot gases and smoke rise because they are lighter; they hit the ceiling and spread out to form a layer at the ceiling. These products of combustion are trapped in the room and continue to affect the growth of the fire, unlike in a free-burning fire that occurs outdoors. Radiation starts to play a greater role in the growth of the room-and-contents fire.

The temperature of the room continues to increase as the fire grows. The room temperature is highest at the ceiling and lowest at the floor level. As the fire intensifies, the visibility will be greatest at the floor level and poorest at the ceiling.

The growth of the fire is limited by the fuel available or by the oxygen available. If the room in which the fire is burning is noncombustible, the only fuel available will be the contents of the room. Likewise, if the doors and windows of the room are closed, a limited amount of oxygen will be present. Either of these conditions may limit the growth of the fire. If flames reach the ceiling, however, they are likely to trigger involvement of the whole room.

PRO TIPS

A burning wastebasket that melts creates a burning pool of fire. This type of fire deposits fuel directly on the floor. In a building with lightweight parallel chord floor trusses, the fire can burn through the floor rapidly and enter the void created by the trusses. This will allow fire to travel freely under the floor. This space is rarely equipped with sprinklers.

Fully Developed Phase

As the fire develops, temperatures increase to the point where the flammable materials in the room are undergoing pyrolysis. Large amounts of volatile gases are being released. If the temperature becomes high enough to ignite the materials in the room, a condition called **flashover** can occur. In flashover, the temperature in the room reaches a point where the combustible contents of the room ignite all at once. This temperature varies depending on the ignition temperature of the room contents. Flashover is the final stage in the process of fire growth.

Flashover is not a specific moment, but rather the transition from a fire that is growing by igniting one type of fuel to another, to a fire where all of the exposed fuel in the room is on fire. If a fire does not have enough ventilation to supply sufficient oxygen, it cannot flash over. If the ventilation openings are too large, the fire may not reach a temperature high enough to bring the whole room to the flashover point. The critical temperature for a flashover to occur is approximately 1000°F (538°C). Once this temperature is reached, all fuels in the room are involved in the fire, including the floor coverings on the floor. Indeed, the temperature at the floor level may be as high as the temperature at the ceiling was before flashover. Firefighters, even with full personal protective equipment (PPE), cannot survive for more than a few seconds in a flashover.

Once the room flashes over, the fire is fully developed. All of the combustible materials are involved in the fire, and the burning fuels are releasing the maximum amount of heat. This condition is sometimes referred to as steady-state burning. The burning gases at the ceiling level radiate heat to combustible materials in the room. Carpets and other objects can ignite from this radiant heat; melting plastics can ignite and drop flaming particles onto materials below them. An average-sized furnished room with an open door will be completely involved in a flash-over within 5 to 10 minutes of the ignition of the fire.

The amount of fire generated depends on the amount of oxygen available. For the fire to generate the maximum amount of heat, ventilation must be adequate to supply the fire with sufficient oxygen. If openings in the room let oxygen in, those same openings will serve as an escape route for hot gases. In this way, the hot gases can spread the fire outside the original fire room. The total fire damage to the room is the result of the intensity of the heat applied and the amount of time the room is exposed to the heat (**FIGURE 3-22**).

Decay Phase

The last phase of a fire is the decay phase (**FIGURE 3-23**). During the decay phase, open-flame burning decreases to the point where there is just smoldering fuel. A large amount of heat build-up means that the room will remain very hot, even though the amount of heat being produced decreases. Although a large amount of heat is available, the amount of fuel available to be pyrolyzed decreases, so a smaller amount of combustible vapors is available. The amount of fuel available dwindles as the supply of fuel becomes exhausted.

A fire in the decay phase will become a smoldering fire and eventually go out when the supply of fuel is exhausted. During this phase, the compartment temperature will start to decrease, though it may remain high for a long period of time. A smoldering fire may continue to produce a large volume of toxic gases and the compartment may remain dangerous even though the fire appears to be under control.

Special Considerations

Flameover

Flameover (also known as **rollover**) is the flaming ignition of hot gases that are layered in a developing room or compartment fire. During the growth phase, the hot gases rise to the top of the room. As these gases cool, they fall and are pulled back into the fire by convection currents. If one of the gases in this mixture reaches its ignition temperature, it will ignite from the radiation from the flaming fire or from direct contact with open flames. In other words, the layer of flammable vapor catches fire. These flames can extend throughout the room at ceiling level. Flames can move at speeds of 10 to 15 degrees (305 to 457 cm) per second. The result of flameover is to increase the temperature in the room. Note that flameover is not the same as flashover, which was described earlier in this chapter.

FIGURE 3-22 The fully developed phase of a typical room-and-contents fire.

FIGURE 3-23 The decay phase of a typical room-and-contents fire.

Thermal Layering

Thermal layering is a property of gases such that the gases rise as they are heated and form layers within a room. The hotter gases are, the lighter they become owing to the decreased density and increased speed of their molecules. As a consequence, the hottest gases will travel by convection currents to the top of the room.

If this normal thermal balance is upset, severe injury can occur. When water is sprayed into the upper part of a room, for example, some of the water will be converted to steam. Steam takes up many times more space than the same amount of liquid water. The steam generated by the application of the water will flood the room and can result in severe burns to firefighters even when they are in full PPE, because the ensuing thermal imbalance replaces the normal layering of gases with superheated steam. The energy in the superheated gases at the top of the room converts the water to steam, and the superheated steam expands and can fill the whole room with steam.

Backdraft

Backdraft is the sudden explosion of heated, oxygen-deprived fire gases when oxygen is reintroduced. The development of a backdraft requires a unique set of conditions. When a fire generates quantities of combustible gases, these gases can become heated above their ignition temperatures. Backdrafts require a "closed box"—that is, a room or building that has a limited supply of oxygen. When the fire chamber has a limited amount of oxygen, the oxygen concentration will be reduced, leading to decreased combustion. If a supply of new oxygen is then introduced into the room, explosive combustion can occur owing to the presence of the superheated gases. This kind of explosive combustion may exert enough force to cause great injury or death to firefighters.

Signs and symptoms of an impending backdraft include the following:

- Any confined fire with a large heat build-up
- Little visible flame from the exterior of the building
- A "living fire," with smoke puffing from the building that looks like it is breathing
- Smoke that seems to be pressurized
- Smoke-stained windows (an indication of a significant fire)
- Turbulent smoke
- Ugly yellowish smoke (containing sulfur compounds)

PRO TIPS

Since the mid 1980s, residential home design, construction materials and methods, and changes in the materials used to manufacture home furnishings have greatly impacted fire development, grow, ventilation, and suppression methods. The impact has been seen with an increase in rapidly developing hostile fires that flashover more quickly than homes built prior to the mid 1980s. The increase in flashover events required investigation, and in recent years, research and experiments conducted by the National Institute of Standards and Technology and Underwriters Laboratory has provided a better understanding of the factors contributing to these events. The result of the research and experiments related to room-and-contents fires concluded that fires in modern homes with modern furnishings can enter a new phase in fire development where the fire becomes dormant or enters the decay stage due to insufficient oxygen to sustain continued grow. This stage results in a ventilation-limited fire that typically has no visible flame or active burning. However, there is still significant fuel from the heat and room content that continues to pyrolize, producing fire gases and smoke. Once outside oxygen is introduced into a decay stage compartment, the result will be a rapid ventilation-induced flashover. As homes and furnishings continue to be built in this manner, more and more homes will have the potential for ventilation-induced flashover fires that will lead to revaluation of the traditional fire growth curve.

Characteristics of Liquid-Fuel Fires

Fires involving liquid or gaseous fuels have some different characteristics from fires that involve solid fuels. To understand the behavior of these fires, you need to understand the characteristics of the combustion of liquid and gaseous fuels.

Recall that solid fuels do not burn in the solid state, but instead must become heated and converted to a vapor combine with sufficient oxygen before they will undergo combustion. Liquids share the same characteristic: they must be converted to a vapor and be mixed in the proper concentration with air before they will burn. Three conditions must be present for a vapor and air mixture to ignite:

- The fuel and air must be present at a concentration within a flammable range.

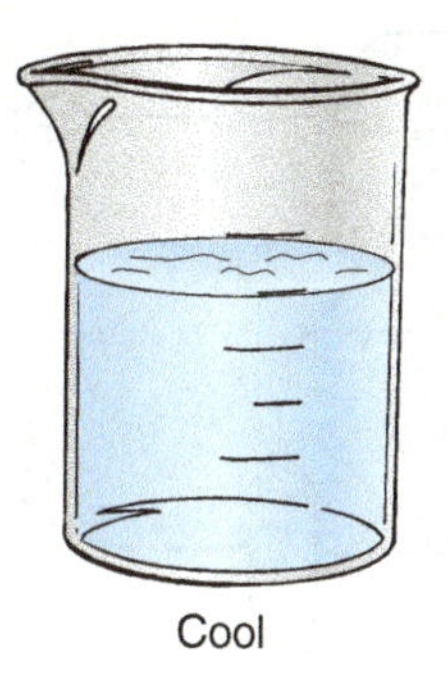

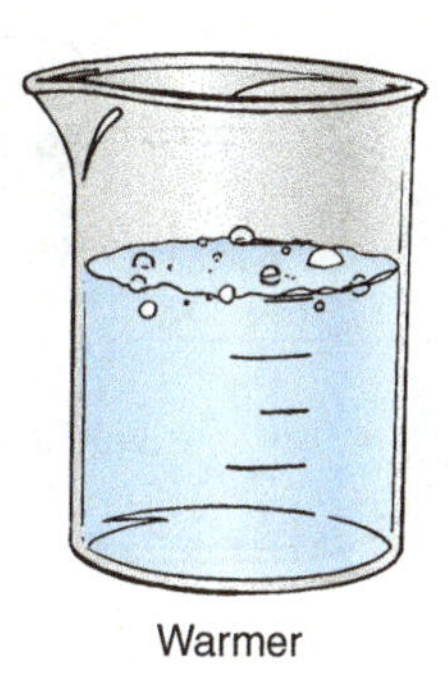

FIGURE 3-24 As liquids are heated, the molecules become more active and the speed of vaporization increases.

- There must be an ignition source with enough energy to start ignition.
- The ignition source and the fuel mixture must make contact for long enough to transfer the energy to the air–fuel mixture.

As liquids are heated, the molecules in the material become more active and the speed of vaporization of the molecules increases. Most liquids will eventually reach their boiling points during a fire. As the boiling point is reached, the amount of flammable vapor generated increases significantly (**FIGURE 3-24**). Because most liquid fuels are a mixture of compounds—for example, gasoline contains approximately 100 different compounds—the fuel does not have a single boiling point. Instead, the compound with the lowest ignition temperature determines the flammability of the mixture.

The amount of liquid that will be vaporized is also related to the **volatility** of the liquid. The higher the temperature, the more liquid that will evaporate. Liquids that have a lower molecular weight will tend to vaporize more readily than liquids with a higher molecular weight. As more of the liquid vaporizes, the mixture may reach a point where enough vapor is present in the air to create a flammable vapor–air mixture.

Two additional terms are used to describe the flammability of liquids: flash point and flame point (**TABLE 3-1**). The **flash point** is the lowest temperature at which a liquid produces a flammable vapor. It is measured by determining the lowest temperature at which a liquid will produce enough vapor to support a small flame for a short period of time (the flame may go out quickly). The **flame point** (or **fire point**) is the lowest temperature at which a liquid produces enough vapor to sustain a continuous fire.

TABLE 3-1 Flash Point and Flame Point

Fuel	Flash Point (°F)	Flash Point (°C)
Gasoline	−45	−43
Ethanol	54	12
Diesel fuel	104–131	40–55
SAE N 10 motor oil	340	171

TABLE 3-2 Vapor Density of Common Gases

Gaseous Substance	Vapor Density
Gasoline	>3.0
Ethanol	1.6
Methane	0.6
Propane	1.6

Characteristics of Gas-Fuel Fires

By learning about the characteristics of flammable gas fuels, first responder inspectors can help to prevent injuries or deaths by identifying potential problems during a fire inspection. Two terms are used to describe the characteristics of flammable vapors: vapor density and flammability limits. For example, during an inspection, an odor of natural gas or other flammable gas or liquid requires investigation to assure there are no leaks, that proper venting exists, and that vapor limits are acceptable to prevent any fire from occurring.

Vapor Density

Vapor density refers to the weight of a gas fuel and measures the weight of the gas compared to air (**TABLE 3-2**). The weight of air is assigned the value of 1. A gas with a vapor density of less than 1 will rise to the top of a confined space or rise in the atmosphere. For example, hydrogen gas, which has a vapor density of 0.07, is a very light gas. A gas with a vapor density greater than 1 is heavier than air and will settle close to the ground. For example, propane gas has a vapor density of 1.51 and will settle to the ground. By comparison, carbon monoxide has a vapor density of 0.97—almost the same as that of air—so it mixes readily with all layers of the air.

Flammability Limits

Mixtures of flammable gases and air will burn only when they are mixed in certain concentrations. If too much fuel is present in the mixture, there will not be enough oxygen to support the combustion process; if too little fuel is present in the mixture, there will not be enough fuel to support the combustion process. The range of gas–air mixtures that will burn varies from one fuel to another. Carbon monoxide will burn when mixed with air in concentrations between 12.5 and 74.0 percent. By contrast, natural gas will burn only when it is mixed with air in concentrations between 4.5 and 15.0 percent.

Flammability limits (or **explosive limits**) are the highest and lowest concentrations for a particular gas in air to be ignitable. These terms are used interchangeably because under most conditions, if the flammable gas–air mixture will not explode, it will not ignite. The **lower explosive limit (LEL)** refers to the minimum amount of gaseous fuel that must be present in a gas–air mixture for the mixture to be flammable or explosive. In the case of carbon monoxide, the lower explosive limit is 12.5 percent. The **upper explosive limit (UEL)** of carbon mon-oxide is 74 percent. Test instruments are available to measure the percentage of fuels in gas–air mixtures and to determine when an emergency scene is safe.

Energy Required for Ignition of Vapors

The principal sources of ignition of flammable liquids include flames; electrical, static, or frictional sparks; and hot surfaces. Flames must be capable of heating the vapor to its ignition temperature in the presence of air in order to be a source of ignition. Electrical sparks from commercial electrical supply installations may be hot enough to ignite flammable mixtures. Hot surfaces can be a source of ignition if they are large enough and hot enough. The smaller the heated surface, the hotter it must be to ignite a mixture. The larger the heated surface in relation to the mixture, the more rapidly ignition will take place and the lower the temperature necessary for ignition.

Boiling Liquid, Expanding-Vapor Explosions

One potentially deathly set of circumstances involving liquid and gaseous fuels is the so-called **boiling-liquid, expanding-vapor explosion (BLEVE)**. A BLEVE occurs when a liquid fuel is stored in a vessel under pressure. The vessel is partly filled with the liquid, and the rest of the vessel is occupied by the same compound in the form of a vapor. A propane tank is an example (**FIGURE 3-25**).

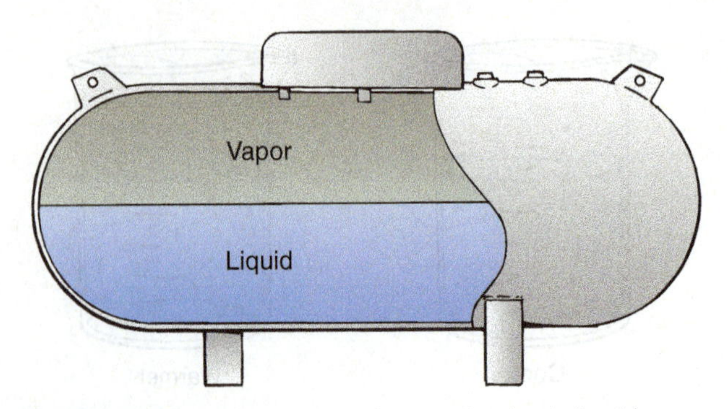

FIGURE 3-25 A propane tank contains both liquid and vapor.

If this sealed container is subjected to heat from a fire, the pressure that builds up from the expansion of the liquid will prevent the liquid from evaporating. Normally evaporation cools the liquid and allows it to maintain its temperature. As heating continues, the temperature inside the vessel may exceed the boiling point of the liquid. The vessel can then fail, releasing all of the heated fuel in a massive explosion. The released fuel instantly becomes vaporized and ignited as a huge fireball.

The key to preventing a BLEVE is to cool the top of the tank, which contains the vapor. This action will prevent the fuel from building up enough pressure to cause a catastrophic rupture of the container.

Fire Growth

Fire growth and spread are greatly influenced by building construction. Whether or not a compartment goes to flashover, for example, is greatly influenced by the materials lining the walls and ceilings of the space. What follows is a description of these hazards and some of the fire protection methods employed to deal with them. It is important for you to be able to identify such potential situations and initiate appropriate actions.

Flame spread, or the more accurate description, fire growth, is a particularly hazardous fire phenomenon. On November 28, 1942, rapid fire growth was responsible for 492 deaths in the Cocoanut Grove nightclub fire in Boston (**FIGURE 3-26**). At the time, flammable decorations were blamed for the rapid spread of the fire. Inadequate exits and overcrowding were blamed for

FIGURE 3-26 Interior design is now regulated by the code.

the monstrous death toll. Pictures from the fire clearly show tell-tale globs of burned adhesives used to glue highly combustible acoustical tile to the ceiling. After other deadly 1940s fires with rapid flame spread in a soldier's hostel in Newfoundland and in Mercy Hospital in Iowa, fire officials began to understand that there was a problem with some building materials.

Building or Contents Hazard

Flame spread, or rapid fire growth, can be a problem caused by both the building itself and its contents. The fire growth building problem can be differentiated by a location characteristic:

- Hidden
- Exposed

The fire growth contents problem can result from:

- Furnishings
- Interior finish, including decorations
- Furnishings
- Building construction elements

In addition to fire growth, high flame-spread materials may contribute heavily to the fire load and to the generation of smoke and toxic products. These are additional threats to the life safety of both occupants and firefighters.

Hidden Building Elements

A wide variety of materials and hidden building elements contribute greatly to rapid fire growth. For example, batt (or paneled) insulation laid in ceilings must be kept free of light fixtures because the heat from the fixture can ignite the paper vapor seal, igniting a fire inside of the ceiling. Any renovation may result in prior materials of construction being covered over by the new materials of construction. It is not uncommon for paneling or wall board to cover combustible insulation or dropped non-combustible ceiling materials and even exposed wiring. A vigilant first responder inspector pays particular attention to these situations.

Combustible fiberboard is commonly used as insulating sheathing on wood frame buildings. It is also used as sound-proofing. This material can support a fire hidden in the walls.

Foamed-plastic insulation is also used as sheathing, concealed in cavity walls, and glued to the interior surface of masonry wall panels. Foamed plastic applied to walls and ceilings for insulation has been involved in many disastrous fires. When such insulation is installed, it should be protected from exposure to flame by a 0.5" (12.7 mm) gypsum board covering. This protects only against ignition from a small source. In a well-developed fire, as the gypsum board fails, the plastic will be involved, possibly suddenly and explosively as the gypsum fails or falls away.

Foamed plastic may be manufactured so that its flame spread is reduced, but it still can melt. It also lacks dimensional stability. When the plastic is used structurally, this may lead to disaster.

Air-duct insulation commonly installed years ago was usually made of a hair felt with a high flame spread. The presently used aluminum-faced foil (not aluminum-faced paper), glass-fiber insulation presents little flame spread problem.

Electrical insulation may be self-extinguishing. However, the tests of this material are conducted on wire not under load. When electrical wiring is operated at or above its rated capacity, the heat can break down the insulation and flammable gases can be emitted. The McCormick Place fire in Chicago is thought to have started in the flaming insulation of an overloaded extension cord. Large groups of electrical wires can support self-sustaining ignition.

Interior Finish

There are three ways in which interior finishes increase a fire hazard:

- They increase fire extension by surface flame spread.
- They generate smoke and toxic gases.
- They may add fuel to the fire, contributing to flashover.

Common interior finishes include such materials as wall-board, wallpaper, lay-in ceiling tile, vinyl wall covering, and interior floor finish items, such as carpeting, in addition to low-density fiberboard and combustible tiles.

In all cases, only approved ceiling, wall and floor finishes are installed in structures and documented to be approved product. In some cases (e.g., healthcare occupancies and educational occupancies), state licensure requirements have prerequisites for interior finishes in buildings. Most materials have a flamespread rating, classification of index.

Ceilings

Both combustible and non-combustible ceiling materials can be found. Older style combustible ceilings that may still be found, as well as wooden ceilings and even wood paneling has been found in older structures. Concrete and drywall/gypsum board construction is prevalent today which provides good limitation of fire growth unless combustible materials are used as cover to the ceiling, similar to walls (e.g., carpeting, tapestry, certain paints and sprayed on products.)

> **PRO TIPS**
>
> Materials used as interior finish must be assigned a flamespread rating. Ratings range from Class A (0 to 25), Class B (26 to 75), and Class C (76 to 200).

Walls

Interior walls may be found of gypsum board, some type of wood, paneling, concrete (usually painted over) or even covered with flammable or combustible products, for example, tapestry, paint, garments. Most non-wood wall assemblies will not quickly contribute to flame spread or fire growth. However, those finishes which have an easily ignitable surface that can sustain fire growth present a problem and require action by the first responder inspector.

Floors

Floor finishes may range from non-combustible products, such as concrete to surfaces to coverings that can sustain fire growth (such as floor tile) to carpeting which can support combustion (wood etc.) due to the use of combustible padding and glue under the carpet as well as the materials of the carpet itself.

Decorations

Present-day decorations and building contents present a major fire hazard. At one time, a popular decoration that provided a real flame spread or rapid fire growth problem was the Christmas tree. Although there are still tragedies during Christmas season, the incidence of such fires has been reduced. Regardless of precautions, the amount of decorations in places of assembly should be strictly limited to reduce the potential volume of fuel for a fire.

Decorations may range from combustible wall hangings to items hanging from automatic fire sprinklers or other ceiling attachments. Being observant is your greatest ally as a first responder inspector. If something looks like it is out of place, it probably is, and an appropriate query and action is justified.

Decorations, usually hanging overhead and/or on the walls are very difficult to control from a fire prevention perspective. There are few applicable regulations. This is typically a generic provision in the code requiring limitations on decorative hangings. This leaves much discretion to the first responder inspector and suggests the first responder inspector act in a judicious and fair manner when finding decorations. Decorations often represent a serious problem because the hazard goes unrecognized even by people who consider themselves fire conscious.

> **PRO TIPS**
>
> The use of carpeting has changed in recent years. It is now also used on walls and ceilings. It is not always evident to designers or even to fire officials that the location where a material is installed may increase the rate at which a fire can grow. A spectacular 1980 fire in the Las Vegas Hilton Hotel spread from floor to floor outside the building because of flammable carpeting on the walls and ceilings of the elevator lobbies. The fire grew into one terrifying fire front extending many stories in height, and claimed eight lives.

It has become commonplace for daycare centers to use carpeting on the walls. Unless the carpeting has achieved the appropriate interior finish fire rating, it must be removed.

Furnishings

Furnishings (drapes, curtains, decorations, chairs, couches, bedding, etc.) all may contribute to fire growth. In general, furnishings (what is in a room or structure) provide fuel for the fire. Their potential to contribute to fire growth is directly related to their materials of construction, their potential to interact with a heat source that can raise its temperature to an ignition point and sustain fire development. Almost any furnishing can burn so fire prevention is the foundation of limiting fire development and fire growth.

Building Construction Elements

Building construction elements may be fire resistive, non-combustible, semi-fire resistive, or combustible. Needless to say, if combustible they have the ability to contribute to the fire load. Depending upon the method of construction, the fire may extend quickly or be contained. Meeting code requirements is important to look for to assure construction limits fire growth or to advise the fire department that full involvement warrants fire department defensive operations.

Flame Spread and Smoke Development Ratings

Flame spread and related smoke development is the movement of fire beyond the burning area and igniting adjacent material, all the while developing products of combustion (smoke) as part of the process. Flame spread and smoke development ratings are based upon testing methodologies established by the American Society for Testing and Materials (ASTM), Test Method E-84. This is commonly known as the "tunnel test." The testing is typically conducted by Underwriters Laboratories or some similar third party testing facility. Each material specification sheet available from the manufacturer has the related rating information on it.

WRAP-UP

CHAPTER SUMMARY

- Matter exists in three states: solid, liquid, and gas.
- Energy exists in many forms, including chemical, mechanical, and electrical.
- Conditions necessary for a fire include fuel, oxygen, heat, and a self-sustaining reaction.
- Fire may be spread by conduction, convection, and radiation.
- The four principal methods of fire extinguishment are cooling the fuel, excluding oxygen, removing the fuel, and interrupting the chemical reaction.
- Fires are categorized as Class A, Class B, Class C, Class D, and Class K. These classes reflect the type of fuel that is burning and the type of hazard that the fire represents.
- Solid-fuel fires develop through four phases: the ignition phase, the growth phase, the fully developed phase, and the decay phase.
- The growth of room-and-contents fires depends on the characteristics of the room and the contents of the room.
- Special considerations related to room-and-contents fires include flameovers, the thermal layering of gases, and backdrafts.
- Liquid-fuel fires require the proper mixture of fuel and air, an ignition source, and contact between the fuel mixture and the ignition.
- The characteristics of gas-fuel fires are different from the characteristics of other types of fires.
- Vapor density reflects the weight of a gas compared to air.
- Flammability limits vary widely for different fuels.
- A boiling-liquid, expanding-vapor explosion (BLEVE) is a catastrophic explosion in a vessel containing a boiling liquid and a vapor.
- Fire growth and spread are greatly influenced by building construction. Whether or not a compartment goes to flash-over, for example, is greatly influenced by the materials lining the walls and ceilings of the space.

KEY TERMS

Backdraft The sudden explosive ignition of fire gases when oxygen is introduced into a superheated space previously deprived of oxygen.

Boiling-liquid, expanding-vapor explosion (BLEVE) An explosion that occurs when a tank containing a volatile liquid is heated.

Chemical energy Energy that is created or released by the combination or decomposition of chemical compounds.

Class A fires Fires involving ordinary combustible materials, such as wood, cloth, paper, rubber, and many plastics.

KEY TERMS CONTINUED

Class B fires Fires involving flammable and combustible liquids, oils, greases, tars, oil-based paints, lacquers, and flammable gases.

Class C fires Fires that involve energized electrical equipment, where the electrical conductivity of the extinguishing media is of importance.

Class D fires Fires involving combustible metals such as magnesium, titanium, zirconium, sodium, and potassium.

Class K fires Fires involving combustible cooking media such as vegetable oils, animal oils, and fats.

Conduction Heat transfer to another body or within a body by direct contact.

Convection Heat transfer by circulation within a medium such as a gas or a liquid.

Decay phase The phase of fire development in which the fire has consumed either the available fuel or oxygen and is starting to die down.

Electrical energy Heat that is produced by electricity.

Endothermic Reactions that absorb heat or require heat to be added.

Exothermic Reactions that result in the release of energy in the form of heat.

Fire A rapid, persistent chemical reaction that releases both heat and light.

Fire tetrahedron A geometric shape used to depict the four components required for a fire to occur: fuel, oxygen, heat, and chemical chain reactions.

Fire triangle A geometric shape used to depict the three components of which a fire is composed: fuel, oxygen, and heat.

Flameover (rollover) A condition in which unburned products of combustion from a fire have accumulated in the ceiling layer of gas to a sufficient concentration (i.e., at or above the lower flammable limit) such that they ignite momentarily.

Flame point (fire point) The lowest temperature at which a substance releases enough vapors to ignite and sustain combustion.

Flammability limits (explosive limits) The upper and lower concentration limits (at a specified temperature and pressure) of a flammable gas or vapor in air that can be ignited, expressed as a percentage of the fuel by volume.

Flashover A condition in which all combustibles in a room or confined space have been heated to the point at which they release vapors that will support combustion, causing all combustibles to ignite simultaneously.

Flash point The minimum temperature at which a liquid or a solid releases sufficient vapor to form an ignitable mixture with the air.

Fuel All combustible materials. The actual material that is being consumed by a fire, allowing the fire to take place.

Fully developed phase The phase of fire development in which the fire is free-burning and consuming much of the fuel.

Gas One of the three phases of matter. A substance that will expand indefinitely and assume the shape of the container that holds it.

Growth phase The phase of fire development in which the fire is spreading beyond the point of origin and beginning to involve other fuels in the immediate area.

Hypoxia A state of inadequate oxygenation of the blood and tissue.

Ignition phase The phase of fire development in which the fire is limited to the immediate point of origin.

Ignition temperature The minimum temperature at which a fuel, when heated, will ignite in air and continue to burn.

Liquid One of the three phases of matter. A nongaseous substance that is composed of molecules that move and flow freely and that assumes the shape of the container that holds it.

Lower explosive limit (LEL) The minimum amount of gaseous fuel that must be present in the air mixture for the mixture to be flammable or explosive.

Matter Made up of atoms and molecules.

Mechanical energy Heat that is created by friction.

Plume The column of hot gases, flames, and smoke that rises above a fire. Also called a convection column, thermal updraft, or thermal column.

Radiation The combined process of emission, transmission, and absorption of energy traveling by electromagnetic wave propagation between a region of higher temperature and a region of lower temperature.

Smoke An airborne particulate product of incomplete combustion that is suspended in gases, vapors, or solid or liquid aerosols.

Solid One of the three phases of matter. A substance that has three dimensions and is firm in substance.

Thermal column A cylindrical area above a fire in which heated air and gases rise and travel upward.

Thermal layering The stratification (heat layers) that occurs in a room as a result of a fire.

Thermal radiation How heat transfers to other objects.

Upper explosive limit (UEL) The maximum amount of gaseous fuel that can be present in the air mixture for the mixture to be flammable or explosive.

Vapor density The weight of an airborne concentration (vapor or gas) as compared to an equal volume of dry air.

Volatility The ability of a substance to produce combustible vapors.

You Are the First Responder Inspector

You are partnering with the fire suppression team of the fire department to assist in creating a prefire plan by inspecting a freight terminal. This freight terminal handles cargo ranging from lumber to large animal feed to manufacturing chemicals. When you arrive at the freight terminal, you are handed a loose leaf book containing Safety Data Sheets (SDS) for all of the potentially hazardous materials that are shipped into and out of the terminal. The bills of laden and shipping labels are also made available to you.

1. **There is a storage room where wooden pallets are stored. A fire in this room would be which class of fire?**
   - **A.** Class A fire
   - **B.** Class C fire
   - **C.** Class K fire
   - **D.** Class B fire
2. **You find an alcove where liquefied petroleum cylinders are being stored. This practice could present a hazard because:**
   - **A.** the tanks could be placed in front of an exit.
   - **B.** if the tanks are involved in overheating, it could result in a BLEVE.
   - **C.** they would be difficult to fill.
   - **D.** the gas would be lighter than air.
3. **You find batteries for material handling equipment being recharged in a small warehouse storage room. This causes generation of hydrogen gas which is explosive. What safety precautions need to be taken?**
   - **A.** Venting above the operation.
   - **B.** Ventilation below the operation.
   - **C.** None. The warehouse vents will take care of the gas.
   - **D.** Fans should be installed to dissipate the gas in the warehouse.
4. **Flashover in a compartment is influenced by which of the following?**
   - **A.** The total BTU content of the material involved
   - **B.** The flamespread of the interior finish and contents
   - **C.** The ambient temperature
   - **D.** The ignition source of the fire

CHAPTER

4

Performing an Inspection

NFPA 1030 THAT INCLUDES CHAPTER 6: FIRST RESPONDER INSPECTOR (NFPA 1031)

First Responder Inspector

- 6.3.5
- 6.5
- 6.5.1
- 6.5.2
- 6.5.3
- 6.5.9
- 6.5.10

ADDITIONAL NFPA STANDARDS

- **NFPA 13** *Standard for the Installation of Sprinkler Systems*
- **NFPA 101** *Life Safety Code*
- **NFPA 220** *Standard on Types of Building Construction*
- **NFPA 520** *Standard on Subterranean Space*
- **NFPA 555** *Guide on Methods for Evaluating Potential for Room Flashover*

KNOWLEDGE OBJECTIVES

After studying this chapter, you will be able to:

1. Describe the types of fire inspections.
2. Describe when the first responder inspector should begin to inspect a building.
3. Describe the pre-inspection process.
4. Describe the fire inspection process.
5. Describe when and how to cite code violations.
6. Describe how to ensure that code violations are corrected.
7. Describe how to investigate a complaint against an occupancy and ensure that the complaint is resolved.

SKILLS OBJECTIVES

1. Perform a fire inspection of a structure.

You Are the First Responder Inspector

You receive a complaint from a building owner regarding a potential fire hazard in the shared employee break room. Upon inspection, you observe a space heater plugged into an extension cord, plugged into a multi-plug adapter. The office manager states that employees have been using the heater this way throughout the winter without any issues. They further state that without the heater, employees would not be able to use the space. However, you notice signs of excessive heat on the extension cord and deformation on the multi-plug adapter. You are a new inspector and even though the situation appears unsafe, you are unsure of what the code requires.

1. What steps should you take now that you are aware of the situation?
2. What can you do immediately to prevent a fire from happening?

Introduction

A fire inspection can reasonably ensure that a building will be safe for the occupants. The first responder inspector's inspection process is performed to verify existing safety features. Fire and life safety systems, and egress elements are operational and being maintained correctly.

Prior to the start of construction, many different building plans will have to be submitted and approved by the various departments of the municipality. The fire inspector and/or the plan reviewer will look at the plans submitted to that they submitted meet the local codes.

PRO TIPS

The first responder inspector typically inspects buildings that have been approved by a Fire Inspector and are occupied. Once the building is occupied the fire inspection process is really just a few steps—be prepared, be professional, be thorough, document, and follow up.

Types of Inspections

There are a number or basic or routine inspections that a first responder inspector must perform. These include annual inspections, re-inspections, complaint inspections, business license or change of occupancy inspections, and self-inspection. **Annual inspections** are inspections where you inspect the building because its turn has come up in the inspection cycle (**FIGURE 4-1**.) Many agencies divide their jurisdiction into smaller inspection areas on a grid. Each area may be assigned a month, so when a month begins, the fire inspectors know that the buildings in that area must be inspected. Inspections may also be performed based on the requirements of NFPA 1730 6.7, or in accordance with a CRR Plan as defined in NFPA 1300.

FIGURE 4-1 Annual inspections are performed yearly.

Reinspections occur when code violations have been noted and you return to see if the owner is now compliant with the code. **Complaint inspections** occur when someone registers a concern of a possible code violation. You must then investigate the complaint to determine if it is valid. The immediacy of a complaint inspection is determined by the type of complaint. A locked exit door should be investigated and corrected immediately, while concerns about a fire extinguisher do not have the same urgency. The first responder inspector typically starts inspecting buildings once they are occupied. The AHJ may assign the first responder inspector complaints to investigate.

Many first responder inspector's inspections evaluate elements including existing sprinkler systems, fire alarm systems, and fire pumps (**FIGURE 4-2**). **Business**

FIGURE 4-2 Construction or final inspections are conducted as a building is being constructed on specific building components including sprinkler systems.

license or **change of occupancy** inspections occur when the building department is notified of a new business requesting permission to open. Prior to issuing a business license or certificate of occupancy, the building department will work with the fire inspection unit to ensure that the occupancy is code compliant.

PRO TIPS

You should have a thorough knowledge of your agency's inspection schedule.

Finally there are **self-inspections**. These are not performed by a fire inspection unit, but by the business owner. Self-inspections are usually performed on smaller buildings or when there are not enough officials to inspect all of the buildings in the jurisdiction. Many fire inspection units ask that a self-inspection form be submitted to the agency.

When Is the Best Time to Inspect?

Once the building is occupied, routine inspections are conducted to determine code compliance, as many circumstances can change after the occupancy is given final approval, such as locked exit doors, stock stored on the floor blocking the means of egress, or improper use of extension cords. Inspections that have been assigned to the first responder inspector should be done at a "reasonable time," typically during normal business hours. Of course, this will vary based on the type of occupancy, but it is often advantageous to observe the business when it is under normal operation.

FIGURE 4-3 Before a fire inspection, you should review any previous fire inspection reports and, correspondence, along with appropriate building department records and permits.

Pre-Inspection Process

Before a fire inspection, you should review any previous fire inspection reports and correspondence, along with appropriate building department records and permits (**FIGURE 4-3**). Most of these items should be found in the building's inspection file and will give you an idea of what to expect prior to walking through the door.

You may also want to review the past occupancy records to develop a list of items that you should be especially aware of during your inspection. This may show a dangerous trend within the business, such as consistently locking the rear exit doors or blocking a means of egress with poor storage practices. Also perform a quick review of local codes and various reference books to refresh your memory on the specific occupancy-related hazards of the building you are going to inspect.

PRO TIPS

Once the building has been approved for occupancy and is open for business, the building plans are filed in the event construction questions about the building come up during routine inspections that cannot be easily answered.

Classification of Occupancy

Occupancy classification describes how the space is actually being used. Occupancy types are identified in the model codes, and the criteria for classifying an occupancy accompany the occupancy classifications. The occupancy identification of a particular building also may be included in previous fire inspection reports or citations. However, if you rely on past fire inspection reports, make sure that the occupancy type has not changed since that report was written. Businesses within a building can change over years, changing the occupancy type. A greeting card store would be a mercantile occupancy, however if the space becomes a restaurant, then the occupancy becomes an assembly occupancy. Checking the building's reports will only tell you what occupancy type was last in the space.

Codes

It is critical to know which codes you can legally enforce in your jurisdiction. The mere fact that there are sets of model codes available does not give you the right to enforce those codes. The local jurisdiction must legally incorporate by reference a specific set of codes. This incorporation is known as **enabling legislation**. It is the law that gives you the right to note any violations of the code and have them corrected.

Once a set of codes has been incorporated, only those codes must be enforced. For example, if the 2006 edition of NFPA 1, *Fire Code*, was incorporated by your jurisdiction, you must use that edition until the ordinance is changed. You would not be able to apply the 2009 edition of NFPA 1 simply because it is newer.

PRO TIPS

Having a copy of the local codes in your office for reference and for a building owner to review is imperative.

PRO TIPS

When checking on overcrowding in a place of assembly, the safety issue at hand trumps the owner's or occupant's wishes. The only way to verify overcrowding is during actual business operation. Advanced notice that there will be routine non-scheduled inspections for overcrowding and exiting related issues may go a long way to minimizing conflicts with owners.

When looking at the model codes, the first chapter is very important as it usually gives the scope of the code, lists the authority and responsibility of the fire inspector, discusses the appeals process, etc. Model codes are applied differently to existing buildings. Generally speaking, unless a building is renovated or modified, building codes are not applied retroactively.

Another item that should be considered is authority to inspect. Some states will require that to inspect public schools you must be certified by the state. There may also be state or local requirements that you must hold a minimum state or national certification to perform fire inspections. If that is the case, there are often requirements that your certification must be renewed every few years, which usually requires showing documentation of continuing education hours.

Tools and Equipment

In preparation for conducting a fire inspection, you should have a set of basic equipment. Additional specialty tools and equipment should be available for shared use by all fire inspectors in the agency. A list of the typical tools you should have access to includes:

- Writing tools—Pens, pencils, markers, and a clipboard to use as a writing surface and for form storage.
- Forms—A sufficient number of the various forms used should be carried, as you will never know what situation you may encounter. In recent years, more departments have been making a move away from paper reports and have been using handheld computers. These computers can allow the inspection to automatically sync up to the database at the department.
- Flashlight—A flashlight is used to check areas behind appliances, rooms where the lighting does not function, and spaces above ceilings (**FIGURE 4-4**).

FIGURE 4-4 A flashlight is used to check areas behind appliances, rooms where the lighting does not function, and spaces above ceilings.

- Measuring device—This can be used to determine exiting concerns.
- Step ladder—Often the business will have shelving or machines which make observing all areas difficult. A small ladder may give you enough extra height to check these areas out.
- Camera—This tool is helpful to document some of the violations that are found. It verifies the violation, and documented examples can be used to train other fire inspectors. Additionally, many departments are taking pictures of all sides of the building and attaching them to the computer files. This is very helpful for the fire department during the preplanning phase. When computers are placed inside vehicles, pictures can be another resource for the incident commander. In addition, if the fire inspection report goes to court, pictures are proof of the condition of the building at the time the picture was taken.

PRO TIPS

Tools and equipment such as sound and light meters should be professionally calibrated by trained technicians.

PRO TIPS

Some tools are not necessary to carry on a daily basis.

Some tools, especially those that are more expensive to purchase or to maintain, such as gas monitors that require periodic calibration, are usually kept in the office for all to use. In most departments, first responder inspectors return to the office on a daily basis making tool sharing quite feasible. However, in a state or county agency, fire inspectors may only return to a central office once or twice a week. This makes coordinating your inspection needs with the other fire inspectors in the office very important.

- Safety gear—Most work sites will require contractors to wear hard hats. Other sites may suggest gloves, goggles, or hearing protection. The fact that you are a fire inspector does not preclude you from following this common sense requirement.
- Electronic devices—Electronic tools may help you communicate, collect data, or reference codes or regulations.
- Coveralls—A layer of durable clothing will protect your work uniform from dust and dirt in areas such as commercial kitchens and woodworking facilities.

Standard Forms

The forms used during an inspection will vary from agency to agency. A set of standard forms should be used by all first responder inspector in the office. Some of the basics forms are:

- Inspection forms or checklists
- Complaint forms

An inspection form is used to note any violations found on the inspection or to note if no violations were found. The inspection form should consist of the following elements:

- A section showing the business name, address, and phone number.
- A section showing the date of the inspection.
- A section showing the area being inspected if the complex is large, if there are multiple buildings, or you are just inspecting one specific section.
- A section for listing any code violations with the specific code citations.
- A section for listing the reinspection date to remind the owner when you will be returning to check on code compliance.
- Areas for the signatures of the fire inspector and a representative of the owner. It is a good idea to also have a place where the names could be printed beneath the signatures.
- A section with a short legal statement stating the first responder inspector's authority for the fire inspection is optional. The legal statement should be reviewed by your agency's legal staff.

There are two primary styles of inspection forms. One form is primarily blank lines on which you must write in a detailed description of any code violations. The second type of form is a checklist (**FIGURE 4-5**). The checklist is an organized way to verify compliance with certain codes and list violations, but this alone may not give the owner enough information to locate and correct problems. Ample space should be provided for you to provide additional details on the code violation. Another downside of a checklist is it encourages the tendency to only look for the items on the list, potentially missing other code violations.

A **complaint form** lists in detail any complaint that is lodged with the fire inspection unit and is investigated. This form should contain the date, time, and

Inspection Checklist
Inspection Procedures

PREINSPECTION CHECKLIST

Equipment:

General

- ❑ Identification (photo ID)
- ❑ Business work hours

Clothing

- ❑ Coveralls
- ❑ Overshoes
- ❑ Boots

Personal Protective Equipment (PPE)

- ❑ Hard hat
- ❑ Safety shoes
- ❑ Safety glasses
- ❑ Gloves
- ❑ Ear protection
- ❑ Respiratory protection

Tools

- ❑ Flashlight
- ❑ Tape measure(s)
- ❑ Pad (graph paper) and pen or pencil
- ❑ Magnifying glass

Test gauges

- ❑ Combustible gas detector
- ❑ Pressure gauges
- ❑ Pitot tube or flow meter

Plans and Reports

- ❑ Previous reports
- ❑ Violation notices
- ❑ Previous surveys
- ❑ Applicable codes and standards

Notes: ______________________________

SITE INSPECTION

Property Name: ______________________________

Address: ______________________________

Occupancy Classification

- ❑ Assembly
- ❑ Educational
- ❑ Day care
- ❑ Health care
- ❑ Ambulatory health care
- ❑ Detention and correctional
- ❑ One- and two-family dwelling
- ❑ Lodging and rooming
- ❑ Hotel/Motel Dormitory
- ❑ Apartment
- ❑ Residential board and care
- ❑ Mercantile
- ❑ Business
- ❑ Industrial
- ❑ Storage
- ❑ Mixed

 (Page 1 of 2)

FIGURE 4-5 A standard inspection form in the checklist format.

Hazard of Contents

- ❑ Light (low)
- ❑ Ordinary (moderate)
- ❑ Extra (high)
- ❑ Mixed
- ❑ Special hazards

Exterior Survey

- ❑ Housekeeping and maintenance

Building construction type

- ❑ Type I (fire resistive)
- ❑ Type II (noncombustible)
- ❑ Type III (ordinary)
- ❑ Type IV (heavy timber)
- ❑ Type V (wood frame)
- ❑ Mixed

Construction problems

Building height ____________ feet ____________ stories

- ❑ Potential exposures
- ❑ Outdoor storage
- ❑ Hydrants

Fire department connection

- ❑ Vehicle access
- ❑ Is it obstructed?
- ❑ Is it identified?
- ❑ Drainage (flammable liquid and contaminated runoff)
- ❑ Fire lanes marked

Building Facilities

- ❑ HVAC systems
- ❑ Electrical systems
- ❑ Gas distribution systems
- ❑ Refuse handling systems
- ❑ Conveyor systems
- ❑ Elevators

Fire Detection and Alarm Systems

See Form A-8.

Fire Suppression Systems

See Form A-10.

Closing Interview

- ❑ Imminent fire safety hazards
- ❑ Maintenance issues
- ❑ Housekeeping issues
- ❑ Overall evaluation

Items to be researched:

- ❑ ________________________________
- ❑ ________________________________
- ❑ ________________________________

Report

- ❑ Draft
- ❑ Review
- ❑ Final

Notes: ________________________________

____

____

____

 (Page 2 of 2)

FIGURE 4-5 *(Continued).*

location of the business; signature spaces; and a record of the alleged violation. When possible, it should also state the name of person making the complaint and the person's contact information; however, the person may remain anonymous. There needs to be a section to indicate what was noted by the inspector when the complaint was investigated, as well as the date and time the inspection was performed. It should also provide a section to advise what corrective action was taken or recommended, and the date the complaint was closed. The first responder inspector is typically assigned a complaint to investigate from a fire inspector.

All of the forms should be in triplicate—one copy for the owner, one for the occupancy file, and the original for the resinspection(s). The original is used to note the date on which the various code violations become compliant. It can also be used to document any conversations you had with the owner regarding the inspection and the related reinspections.

SAFETY TIPS

Fires can grow due to many hazardous conditions. While fires can start from many things, fire growth often times is related to housekeeping issues. Some of the more common issues include:

- Storing stock too close to the sprinkler heads
- Garbage cans overflowing
- An accumulation of materials in furnace and water heater rooms
- Storing boxes and mops too close to pilot lights on furnaces and water heaters
- Storing flammable materials incorrectly

Scheduling and Introductions

There are two ways to begin the inspection—scheduled and unannounced. Each has its pluses and minuses. Scheduled inspections ensure that you will be able to fully conduct the fire inspection; however, it also allows the business to conduct its own pre-inspection and correct any usage violations prior to your arrival. The unannounced inspection allows you to get a "real" look at the business and understand how the business operates daily. A drawback to this inspection is the risk of being turned away due to an inability of the business to accommodate you at the specific time, wasting your trip.

While the codes give you the authority to conduct fire inspections, you are not allowed to do so without permission of the owner. Performing a fire inspection without permission is trespassing. The exception to this is an **exigent circumstance**, meaning an immediate life safety issue which requires immediate actions to be taken. Non-operating emergency lights or fire extinguishers with inspection tags out of date are not exigent circumstances. Locking emergency exits in a mall during the Christmas season is an exigent circumstance.

If permission to conduct the fire inspection is denied multiple times without valid explanation, then, as a last resort, you should request a court order and an inspection warrant to conduct the fire inspection. At times, working with the municipality provides the owner motivation to allow the fire inspection because the municipality often issues business licenses and often require compliance with the local codes.

Upon arrival, you must first seek permission from the owner, manager, or other individual with the authority to allow you to inspect the occupancy. A teenager working as a cashier may not have the authority to grant permission. Obtaining permission can be as simple as walking in to the business and asking to speak to the owner or manger. Often, the owner or manager will accompany you during the inspection. If not, he or she will typically assign another individual to assist you. It is always best to have a representative of the business along with you during the inspection. He or she can quickly answer any questions and can give you access to locked areas. Having a representative witness your inspection also lessens the possibility of you being falsely accused of an action.

PRO TIPS

Inspecting some occupancies during your normal business hours may not provide a true picture of what the business is like when fully operational. This is especially true of assembly occupancies such as nightclubs. Performing a full inspection of a nightclub at 11:00 PM might be difficult so it would be more prudent to conduct the full inspection during the day. Inspections should not disrupt business more than is absolutely necessary. This will allow for time to conduct a detailed inspection without disturbing the business operation; however, it may still be necessary to do a spot check during their normal business hours. Those spot checks should focus on occupant loads, visible exits, that the unlocked exit doors, and the clear means of egress.

PRO TIPS

With few exceptions, inspection timetables are subject to each agency's policy. Some may state once a year, with target hazards being inspected twice a year. Local ordinances may specify the frequency of inspections. Additionally, occupancies such as schools, hospitals, and daycare centers may, through state law, require more frequent inspections. The best plan is to go through each building once a year, or in accordance with NFPA 1730 or a CRR plan in accordance with NFPA 1300.

Asking for permission, as opposed to demanding to make an inspection, will help develop good rapport with the owner. When permission is not granted, most often the occupancy will simply ask if you can come back at a time more convenient for them. This should be honored.

The Fire Inspection Process

The process of inspecting a building will vary with the occupancy. Occupancies vary in size and scope. Although some will be single-story buildings, others may have multiple buildings or multiple floors. Before the onsite inspection, it is important to verify that you have current information about the occupancy. Check your forms to see if there are any documents you will need to request of the owner. If your forms show a specific occupancy type, you should confirm that the occupancy class has not changed.

Presentation

First impressions count! Your first few minutes with the property or occupancy representative will set the stage for your entire professional relationship with that person. You will be judged on your appearance, your attitude toward your work, and the way you interact with others (**FIGURE 4-6**). Wearing a uniform generally makes it easy to see what agency you are representing. In spite of the ease of recognition a uniform brings, many agencies require the first responder inspector to have an ID card visible. You should never hesitate to show any form of identification when asked and should even compliment the person for asking. This shows that you realize that others may misrepresent themselves for unscrupulous reasons. It also shows that you approach your job as a first responder inspector seriously and professionally.

FIGURE 4-6 You will be judged first upon your appearance, your attitude toward your work, and towards those with whom you interact.

If your jurisdiction does not have uniforms, consider wearing appropriate professional attire. In many cases this could be a dress shirt, tie, badge, and nametag. During hot summer months, a short sleeve shirt with a collar may be appropriate. Having the name of the organization you represent on the shirt when possible, will connect you with an agency. When that is not possible, ID cards become much more important.

When working at job sites, safety concerns change dress code expectations. Boots and jeans may be appropriate. It should not be forgotten that when at a job site most workers are required to wear hard hats. You are not immune to hard hat or eye protection requirements.

Once you have arrived with the appropriate uniform, show the appropriate professional attitude and mannerisms. You can show professionalism by waiting your turn to speak and thanking people for their assistance. Do not smoke, chew gum, make loud noises, or walk around while waiting for the owner. Your mannerisms outwardly demonstrate your mental attitude. For instance, when first meeting with the property or occupancy representative, make eye contact and firmly shake hands. Introduce yourself by stating your name, title, and the reason you are there—to seek permission to inspect the property. Do not forget that you represent the fire inspection unit and the fire department.

In the event that the owner shows some hostility towards you, remain professional. Ask the owner if another first responder inspector or fire inspector can perform the inspection. When you return to the office, make sure your supervisor is notified of the problem. If there has been a pattern of this, it may be appropriate to ask another official to accompany you on the next inspection.

Conducting the Exterior Inspection

Prior to actually meeting the building owner, certain observations can be made regarding the building (**FIGURE 4-7**). From the exterior you can observe vehicle access, fire lanes, fire hydrant access, caps missing on fire hydrants, sprinkler connection visibility and access, and exterior building issues. These include broken windows, tilting walls, and missing or falling bricks (**FIGURE 4-8**). Other things to note would be ponds, gates, and fences or other barriers. If this is a large complex it may be helpful if you have a site plan of the property.

FIGURE 4-7 As you drive up, take note of the exterior of the building.

FIGURE 4-8 From the exterior you can observe vehicle access, fire lanes, fire hydrant access, or caps missing on fire hydrants.

Before conducting an exterior inspection, notify building personnel of your presence and intentions. This is more professional than explaining your presence and behavior to security guards or other personnel.

Conducting the Interior Inspection

Adequate time should be allotted to perform a thorough inspection of the building's interior and complete required documentation. If this is your first time at the building, additional time should be allotted to learn the building layout. If you have inspected the building before, it may be a good idea to rotate other first responder inspector through the buildings. Different eyes see different things.

When conducting the fire inspection, follow a predetermined and structured order so that no areas of the building are missed. You may begin at the top floor and work down, or start at the front and work toward the rear, or start by going left, or right, and continuing that direction until you arrive back at the starting point. It is important that the inspection be systematic, thorough, and well-documented. It is important to look into every room.

In the cases of large, complex occupancies and properties, you may need to break the inspection into sections or buildings in order to make sure nothing is overlooked. This may entail conducting the fire inspection over multiple days. You must also remember that the person who is accompanying you has other job duties that are not being performed while he or she is escorting you.

Having the owner along helps you gain access to locked or restricted areas. Just because a door is locked does not mean that it is exempt from the inspection. There may be restricted areas that require signing into a log in order to access them. In rare cases there may be areas that contain trade secrets of the business and entry may be refused. Documenting those areas where access is not fully gained is important. Rapport with the owner is crucial. Try to determine information by asking the following:

- "How large is the area?"
- "Can I just step inside the door for a quick once over?"
- "Could someone from a safety committee provide some documentation of the hazards?"

Ask questions about the building during the inspection. The answers may give some insight to some possible hazards you have not previously noticed. Be thorough and do not rush. Often the owner will escort you to where he or she thinks you want to go. Do not be afraid to stop and look at areas along the way.

Code Violations

When documenting code violations, it helps to be specific about the problem and the needed steps toward code compliance. Including reference to the specific code and section that is in violation ensures that the issue is a violation of a specific requirement and helps the owner/occupant to better understand the requirements of the code. The owner will need to fix the violation, and you or another inspector must be able to relocate the problem area when it is time for reinspection. Rather than simply stating the issue, you should indicate what is necessary in order to reach compliance. For example, if a fire extinguisher is out of date, state that the fire extinguisher requires a current inspection tag. This provides the owner with direction, increasing the likelihood of compliance during reinspection.

On rare occasions, you may encounter a process or hazard that is extremely dangerous to the occupants. This is considered an exigent circumstance and may warrant closing a business. This action needs to be considered with extreme care. Some fire inspectors feel that have a right to close a business if violations are not corrected. Businesses can be closed, but unless it is an exigent circumstance, it must be done through legal means. In most cases, the first responder inspector would document the findings and report to their supervisor, such as the fire inspector. Legal action may be pursued at the direction of the supervisor, and at this point, the documentation provided by the first responder inspector is crucial. Accurate, factual information that is reported by the first responder inspector is vital if legal action takes place.

Fire Protection Features

Fire alarm systems and automatic fire sprinkler systems should be a top priority when evaluating the building's fire protection features (**FIGURE 4-9**). In addition, fire hydrants on or close to the property should be routinely inspected to ensure working order. If the department or municipal water purveyor inspects the hydrants regularly, you may only want to check that the caps are in place and that they are not stuck or painted closed.

FIGURE 4-9 Fire alarm systems and automatic fire sprinkler systems should be at the top of your list when taking into account the building's fire protection features.

Fire detection systems can detect the presence of smoke and fire, alert occupants, notify the fire department, activate fire suppression systems, close fire doors, open smoke vents, and control the building's heating, ventilation, and air conditioning (HVAC) systems. You should have a working knowledge of how these components perform. All of the features of a fire detection system must be inspected; however, it should not be your responsibility to do more than a visual inspection of the fire detection system components or verifying maintenance records. Besides being very time consuming, you could be held liable if a piece breaks or the system cannot be reset. You should note on the inspection form that a recent full fire detection system inspection report must be submitted by the owner showing what devices were tested and if they passed inspection.

It is important to note which type of devices are installed. If heat detectors are installed where smoke detectors should be installed per the code, this is a violation of the code. A smoke detector should be located in an area best suited to smoke detection, for example, in a foyer, lobby, or conference room. A smoke detector located in a machine shop or garage may not achieve the same intended goal and will cause unintended alarms.

The same logic holds true for a fire sprinkler system. While there are typically not many parts that need to be tested, a sprinkler inspection report should be provided by the owner. You are obligated to check for closed valves, proper pressures on various gauges, missing sprinkler coverage, and that heads are installed properly.

A kitchen may call for a specialized fire suppression system (**FIGURE 4-10**). As with other fire suppression systems, you should witness the initial installation to be certain that gas valves close, electricity and fans shut off, and all other components of the system function properly. Following the final inspection, a minimum of an annual inspection should occur. A visual inspection of the fire suppression system during the fire inspection would include looking for caps on the nozzles, accumulations of grease, missing nozzle coverage, and grease filters turned the wrong direction.

FIGURE 4-10 A kitchen may call for a specialized fire suppression system.

© Aprescindere/Dreamstime.com

Hazard Recognition

The role of the first responder inspector is to ensure a safe building for all occupants. During the fire inspection you should be on the lookout for various hazards. There are a few hazard violations that routinely appear, including:

- **Electrical**—Electrical cords cannot be spliced; circuit breakers must be identified; extension cords cannot be used in place of permanent wiring, openings are not allowed in electrical panels; clear access to the electric panel must be maintained; cover plates are needed for junction boxes, switches, and outlets; and no multi-plug adapters are allowed.
- **Exit/emergency lights**—Lights must function properly and must not be obstructed.
- **Exiting**—Exit doors must be operational and unobstructed, doors must close and latch but may not use deadbolt locks, storage is not allowed in halls or stairwells, exit doors must swing outward, and exit signage is required.
- **Fire extinguishers**—Fire extinguishers must have current inspection tags and a minimum of 2A10BC rating; extinguishers must be mounted properly, unobstructed, and operational, indicated by proper signage where appropriate; extinguishers should be properly spaced for minimal travel distance; they should be of the proper type for the hazard; and there should be enough extinguishers to comply with fire codes.
- **Fire detection and suppression systems**—This equipment must be accessible and kept in a normal status, fire department connections must be capped and accessible, storage cannot be too close to sprinklers, and rooms must be properly labeled. Specialized annual inspections are required on these systems.
- **General**—A key box containg proper keys may be required, good housekeeping must be maintained, address numbers on the building must be visible, high pressure cylinders must be secured to the wall, gas meters must be protected, fire hydrants must be visible and accessible, no excessive amounts of flammable liquids may be stored, flammable materials must be kept in the proper containers, ashtrays should be provided in smoking areas, no smoking signs must be provided where smoking is not permitted, emergency vehicle access must be unobstructed, and fire lanes must be identified.

Contents

It makes no sense to inspect a building but not inspect its contents. On January 16, 1967, Chicago's McCormick Place, a large convention center thought to be fire proof due to its steel and concrete construction, burned to the ground. The fire protection needs to match the hazards within. The size or construction of a building is not the determining factor when classifying a content hazard. A 50,000 square foot (4645 m^2) building with storage of aluminum canoes is less hazardous than a 5000 square foot (465 m^2) building storing plastic cups. This is because the amount of heat released with plastic cups is significantly greater than the canoe.

Building Features

Means of egress, fire doors, fire walls, staircase enclosures, smoke ventilation systems, hung ceiling systems, emergency lighting, and exit lighting should be inspected (**FIGURE 4-11**). Look for the following items:

- Check the means of egress including stairway enclosures.
- Make certain that doors swing in the right direction.
- Travel distances and the common paths of travel are not exceeded.
- Doors close and latch.
- Exit doors are not locked.
- Fire doors should close and latch. If there are automatic devices used to operate and release the door, those pieces must function smoothly.
- Smoke evacuation systems and systems designed to pressurize a stairwell or floor are complicated and should be tested by a professional company.

FIGURE 4-11 Means of egress must be inspected.

© Jones & Bartlett Learning

- Any missing ceiling tiles.
- Exit lighting requires looking to see if the lights are illuminated.
- Emergency lighting will require some type of testing. In most cases this can be as easy as pushing the test button to see if the lights illuminate. It gets more difficult when those lights are more than 15 ft (457 cm) off the floor. Often times those units will have a separate circuit breaker and the lights will come on simply by turning it off. Never do that yourself. Let the building owner do that, as you will have no idea what else may be tied into the breaker. There have been instances where a breaker was turned off and computers also shut down, causing the loss of many hours of work.
- More mundane features such as carpeting and interior finishes must also be checked for compliance. In certain public buildings, the furniture needs to have a tag indicating that it will not promote flame spread.

Many areas of a building are obscured from normal view and you may need to enter concealed spaces, voids, and areas above the hung ceiling and the true ceiling of an occupied space. Visual inspections ensure the existence and integrity of fire protection features (**FIGURE 4-12**). Since these are void areas, the prime reason to look behind these areas is to ensure that there are no breaches in the walls or ceilings to ensure that a fire cannot pass into another area. Look for adequate fire stopping of penetrations and that walls are tight to ceilings.

FIGURE 4-12 Buildings built long before the first responder inspector was born may be renovated multiple times hiding a multitude of construction and fire protection problems. Checking for vertical openings and hidden shafts is a must in older buildings.

© M. Niebuhr/Shutterstock

Interior Finish

As a first responder inspector, it is important to understand the impact various interior finishes have on the potential to help or hinder fire's movement. The interior finish is the exposed surface of the floor, walls, and ceiling. Anything attached to them will help to control the speed at which fire would spread. Drapery, curtains, and the like would not be counted as an interior finish unless it is attached to the wall. As building plans are submitted there should be listing of what the interior finishes are and rated for in the various areas within the building. The model code specify the interior flame spread rating for various buildings. Class A is the highest rating and has a flame spread of 0 to 25. The other classes are B and C, and the flame spread goes to 200. After that, there is no rating given.

As you conduct your inspection, evaluating the interior finish includes examining finishes on walls, floors, and ceilings. It is important to ensure that combustible interior finishes (e.g., plywood paneling, cloth, or decorative wood) meet code requirements. This is determined initially through the plan review process and confirmed during inspections prior to occupancy of the building.

Some occupancies may have stage curtains or a foam type product on the wall. Copies of the certificates

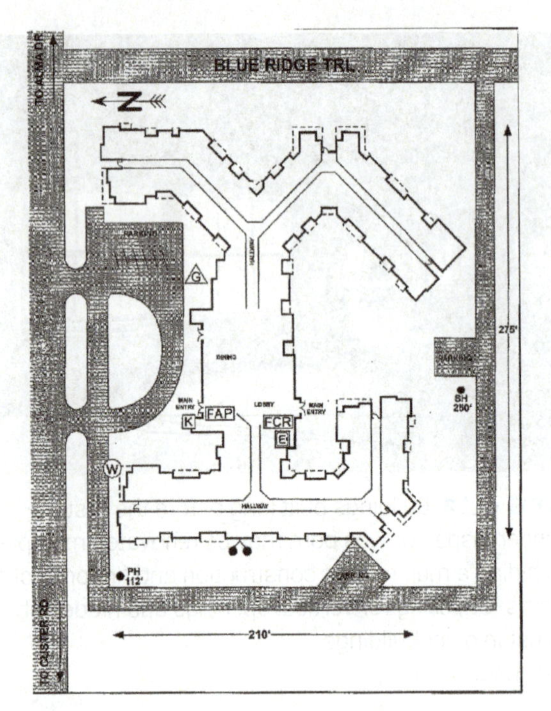

FIGURE 4-13 A sample preplan sketch.

indicating that the product is fire rated and approved for its intended use are good to place in inspection files. These certificates may also list a flameproofing material that must be reapplied every number of years. Carpets placed on walls lose their flame rating unless they have been tested and listed for wall use.

Preplan Sketch

Often times a preplan sketch for the fire department will be drawn during the inspection process. This process takes time, so it may be easier to conduct the inspection and return at a later date for the preplan sketch (**FIGURE 4-13**). If that is the case, when asking permission to walk the building for the preplan sketch, make it clear that you are not there to conduct a fire inspection, but just to draw a floor plan for the fire department's use in case of an emergency. Also consider taking pictures of the property and obtaining a building plan.

Post-Inspection Meeting

It is important to have a conversation with the owner regarding the findings of the inspection (**FIGURE 4-14**). You must always take the high road. It is common that many building owners do not like to see the first responder inspector show up. To them, the first responder inspector may just be someone that costs them money. When talking with the owner about the inspection, be firm in your needs, while at the same time listen and be

FIGURE 4-14 It is important to have a conversation with the owner regarding the findings of the inspection.

empathetic to his or her concerns. When there are disagreements, understanding the owner's viewpoint can often create a middle ground for compliance.

Hopefully the owner has accompanied you on the inspection and items noted were explained at that time. If not, you may need to take the owner to some locations to explain the code violations noted. It is important that you fully explain the reason behind any code violations. To just say that "because the code says so" will not help your cause. If the owner can be educated about the hazards, hopefully their reoccurrence will be lessened.

You should also take this time to note any other concerns that may not be actual code violations, but would be of interest to the owner such as fire extinguishers due for inspection in two months, making sure the sidewalks are kept clear of snow, etc. This can be noted on the inspection form for documentation. A copy of the inspection form should be left with the owner. If there are code violations, you should state when you will return to reinspect. As a final act, thank the building occupant or representative for their cooperation, even if it was lacking. This shows your professional attitude.

> **PRO TIPS**
>
> You should be aware of your local appeal process and inform the owner of his or her right to appeal.

Documentation

Documenting the inspection and code violations in writing and in a prescribed format is the best way to attest to the findings of your inspection. Legal requirements and lawsuits have changed the way the fire inspection unit must document each and every element of their inspections. Inspection records could be recalled many years after an inspection has occurred. Two prime reasons for that would be a lawsuit or if there were a fire and people are looking to see what was noted in previous years' inspections.

The inspection is documented in the inspection report. The inspection report can be a check list, free form, or a combination of both. How to write an inspection report is covered in detail in *Chapter 10: Writing Reports and Keeping Records.* When there are a great many issues or the issues are complex, it may be best to forgo the standard inspection form and to create a formal letter documenting the violations and the expectations. Any non-compliance or code violation noticed during the inspection must be documented in the written report of the inspection. A copy of the inspection form and a date for a return inspection should be left with the building owner. If necessary, mailing the form is acceptable; however, the only way to document if it was received is by certified mail.

Noting the date for the reinspection is important because by documenting a violation, you are stating you are aware of fire or life safety code violations. In Adams v. State of Alaska, October 1, 1976, the State undertook a fire inspection in a hotel and noted hazards, some of which were extreme life hazards. The State advised the hotel of the hazards, but did not take additional action. Several months later, a fire killed and injured many people. As a result of the inspection, the State owed a duty to exercise reasonable care to abate the hazards. By failing to return and have the hazards abated, the State breached their duty by their inaction.

Generally, the inspection report is best delivered in person. This way the owner may ask questions. A few extra minutes explaining why an item is a code violation provides guidance and understanding to the owner, and may lessen the chance of its reoccurrence. Issues that may seem clear to you may be unfamiliar to the owner, so extra time must be spent to explain those issues. When appropriate, ask if the owner would like a copy of the section of the code noting the violation. Upon returning to the office, simply copy that section of the code and fax or email it to the owner.

At no time should you become the agent of the owner. Only offer guidance based on the code. It is not your job to fix the code violation. Attempting to do so could lead to you assuming responsibility for the fix, which could become legally problematic.

When documenting an inspection, keep in mind that the information must be easily retrievable and easily readable by other fire inspectors and the occupancy owner. An inspection report stored on your portable computing device is great for you but is of no value to anyone else.

Code Violations

Code violations are not always black and white in nature. There is the letter of the code and the intent of the code, and you should consider both. Much of this wisdom comes with experience. For example, if the code states a fire extinguisher cannot be mounted more than five feet from the floor, and you inspect a well-kept building with an owner conscientious of safety concerns; however, you note the fire extinguisher is six feet from the ground and has been located there for a number of years. Is this a violation of the code or have they met the intent of the code? How about a small building where the only exit signage you find is a sign above the front door that is not internally illuminated? These types of issues are the judgment calls an inspector must routinely make. Would you be completely in the right if you chose to cite those as code violations? Absolutely.

All noted code violations must be corrected as soon as possible. Some items, such as locked exit doors, must be corrected immediately; others, such as unlit exit signs, do not have the same degree of urgency. If you note violations that have been overlooked for years, the owner may question the violation. This can be a tough situation. The only thing that can really be said is that for some reason this was missed on previous inspections but must now be changed to comply with the code.

When an owner or occupant receives an inspection report, it constitutes an important part of their business and business protection. The purpose of the inspection and the subsequent report is compliance with applicable codes. Follow-up dates are necessary when code violations have been found, providing the timetable for corrective action on the owner or occupant's part. Code violations should have reasonable timetables for compliance. Remember: what is reasonable to you may not be reasonable to the person receiving the report. When assigning the correction time to each violation, consider many factors including the seriousness of the violation, financial cost to correct the violation, and the ability to get the violation corrected. Replacing light bulbs in exit signs is not difficult or expensive, but adding a sprinkler system is much more involved, not to mention more costly.

Any code violations that are corrected during the inspection must also be noted in the final report. Document the fact that a code violation was actually noted during the inspection, by marking the violation with "complied on-site." Keep in mind that easy-to-fix code violations generally reappear once you leave.

When there are complex code violations, more research into the appropriate codes and standards may be necessary. In this situation, advise the owner that you are not certain about a specific item and you want to make certain of the code requirement before finalizing the report. You would not want to cite a violation when one does not exist. For example, if you encounter a chemical and are not certain how much can be stored, if it must be separated from other chemicals, or if it should be in its own room, referencing the code may be required.

When noting code violations, citing the code reference adds legitimacy to your assessment and allows the owner to look up the specific area themselves, should he or she choose. The vast majority of the time, the owner will not question the violation and will be more concerned about time frames or costs. For those owners who question a violation, you should be able to furnish the reference quickly if you elected to not cite code directly. That may mean having a master list of common violations with the associated code reference(s) or knowing where to look in the codes for the answer.

Reinspection dates are at your discretion and agency policy. Many fire inspection units routinely give 30 days prior to their return; however, some agencies have found that shorter time frames, such as two weeks, keep the violations more in the fore-front of the owners' priorities. If items are severe a day or two may be appropriate. Large ticket items it may require months or years to complete, but more frequent follow-ups will ensure that the owner is taking some action, such as getting bids for the work.

The reinspection date is not when the code violations must be complied; instead, code violations must be corrected as soon as possible. The reinspection date is just an approximation of when you will return to check on code compliance. If you are allowing multiple dates for compliance, the inspection form should note that.

As a last resort, if there is non-compliance following repeated efforts, the municipality may need to be called in. They typically have the leverage of issuing fines and revoking business licenses when appropriate. When all else fails, going to court may be needed. It is unfortunate and time consuming, but it shows the seriousness of the inspection process. If after a legal proceeding, a violation is allowed to exist, the judge is the one that has allowed it, not the inspector.

PRO TIPS

When returning to check on compliance, only look for those items in question.

PRO TIPS

Save pertinent notes from an inspection as part of the file. Don't throw away something today that you feel is irrelevant. In the future, you may be wish you had those scribbled notes to clarify something you have put in your formal report.

PRO TIPS

Make sure you know the occupancy classification of what you are inspecting, and that you are using the correct edition of the appropriate code as a reference. Just because a new code had been adopted, it does not automatically mean existing occupancies must meet the new requirements.

Investigating Assigned Complaints

When the fire inspection is the result of a complaint, you must have the information regarding the specific problem that needs to be addressed. A complaint that an occupancy is "dangerous" provides little direction to the inspector. Whenever possible, gather additional information about the nature of the risk, such as exit issues, overcrowding, or a specific dangerous situation. While a "fire trap" is not very specific, the complaint must be evaluated. How soon depends on the urgency of the complaint.

When responding to a complaint, the owner should not know that you plan to inspect because it is important that you see the condition as it exists, not after there has been an opportunity to repair it. Once the complaint is investigated, the building owner should be advised of the results. If there are code violations, a time frame should be given for compliance. Some items, such as locked exit doors, should be corrected prior to your leaving. Once on-site for the investigation, additional problems may be found, necessitating a complete fire inspection in the immediate future.

WRAP-UP

CHAPTER SUMMARY

- A fire inspection will reasonably ensure that a building will be safe for the occupants.
- There are a number or basic or routine inspections including annual inspections, reinspections, complaint inspections, business license or change of occupancy, and self-inspection.
- It is critical to know which codes you can legally enforce in your jurisdiction. The mere fact that there are sets of model codes available does not give you the right to enforce those codes. The local jurisdiction must legally adopt a specific set of codes.
- The forms used during an inspection will vary from agency to agency. A set of standard forms should be used by all fire inspectors in the office. Some of the basics forms are:
  - Inspection form
  - Final or construction inspection form
  - Complaint form
  - **Stop work order**
- While the codes give you the authority to conduct fire inspections, you are not allowed to do so without permission of the owner. Performing a fire inspection without permission is trespassing. The exception to this is exigent circumstances.
- The process of inspecting a building will vary with the occupancy. Some occupancies may have multiple buildings, others will be multiple floors, others will be very small buildings, and another could be a massive one story building.
- The exterior inspection includes vehicle access, fire lanes, fire hydrant access, caps missing on fire hydrants, sprinkler connection visibility and access, and exterior building issues. These include broken windows, tilting walls, and missing or falling bricks.
- When conducting the interior inspection, follow a predetermined and structured order. You may begin at the top floor and work down, or start at the front and work toward the rear, or start by going left, or right, and continuing that direction until you arrive back at the starting point. It is important to look into every room.
- Often times a preplan sketch for the fire department will be drawn during the inspection process. This process takes time, so it may be easier to conduct the inspection and return at a later date for the preplan sketch.
- A post-inspection meeting should be the last step in the physical inspection of the building. When talking with the owner about the inspection, be firm in your needs, while at the same time listen and be empathetic to his or her concerns. When there are disagreements, understanding the owner's viewpoint can allow for a middle ground for compliance.
- Documenting the inspection and code violations in writing and in a prescribed format is the best way to attest to the findings of your inspection. Inspection records could be recalled many years after an inspection has occurred. Two prime reasons for that would be a lawsuit or if there were a fire and people are looking to see what was noted in previous years' inspections.
- Code violations are not always black and white in nature. There is the letter of the code and the intent of the code, and you should consider both.
- When the fire inspection is the result of a complaint, you must have the information regarding the specific problem that needs to be addressed.

KEY TERMS

Annual inspections Inspections performed as part of the regular inspection cycle

Business license or change of occupancy inspections Inspections that occur when the building department is notified of a new business requesting permission to open

Complaint form Form that lists in detail any complaint that is lodged with the fire inspection agency and is investigated

Complaint inspections Inspections that occur when someone registers a concern of a possible code violation

Enabling legislation Legislation in which local jurisdiction adopt a specific set of codes

Exigent circumstance An immediate life safety issue which requires that immediate actions be taken

KEY TERMS CONTINUED

Reinspection An inspection performed to determine if code violations have been corrected

Self inspections Inspection performed by the building owner or occupant

Stop work order A form used when contractors do not have the clearance for performing the work, or when work must be corrected prior to performing additional work

You Are the First Responder Inspector

There is a commercial building has been renovated in your area. During shift, you notice that it is now open and occupied. It is now your responsibility to conduct the occupancy inspection required by the fire inspection unit. The inspection has been assigned to you, and it is scheduled for today. While the business has relocated here, it is similar to the prior business that occupied the building, they manufacture children's toys.

1. **What do you plan to do in preparation for making the initial inspection?**
   - **A.** Visit the property and look around on your off hours.
   - **B.** Look in the file of the toy company's current location for past history of violations and the degree of compliance.
   - **C.** Go to the property and introduce yourself as soon as they take occupancy.
   - **D.** Talk to the building inspector to make sure everything is code compliant so you don't have to walk the entire building.
2. **Before you get a request to inspect the building, the owner invites you to visit the old location and meet with him to discuss code compliance in the new building.**
   - **A.** Don't go, this would be a warrantless search and therefore illegal as you do not have an inspection request nor court order.
   - **B.** Meet with the owner and tour the building to find out what you will have to address when they move.
   - **C.** This would be a wasted trip as the new building would be code compliant as it is brand new.
   - **D.** Meet with the owner and immediately put him on notice that you will not tolerate violating your fire code.
3. **You learn that manufacturing toys requires vast amounts of different kinds of chemicals, some of which react with each other. How do you feel is the best way to handle this?**
   - **A.** Take a chemistry course at the local community college.
   - **B.** Talk to other inspectors and see what they know about chemistry.
   - **C.** Learn who the chemical supplier is and ask for assistance.
   - **D.** Allow the owner to do as he feels is right; he has been in business before.
4. **It appears that this inspection is going to be very complex and confusing, so the best format to use for the inspection report would be:**
   - **A.** a check off sheet with items and check boxes.
   - **B.** a verbal interview with the owner stating what is required with nothing in writing that can be used against the inspector.
   - **C.** a detailed report in essay form with attachments received from suppliers.
   - **D.** all that is needed is to refer the owner to comply with the code and set a date for compliance.

CHAPTER

5

Occupancy Safety

NFPA 1030 THAT INCLUDES CHAPTER 6: FIRST RESPONDER INSPECTOR (NFPA 1031)

First Responder Inspector

- 6.5.5
- 6.5.4
- 6.5.9

ADDITIONAL NFPA STANDARDS

- **NFPA 1** *Fire Code*
- **NFPA 13** *Standard for the Installation of Sprinkler Systems*
- **NFPA 70** *National Electric Code*
- **NFPA 72** *National Fire Alarm and Signaling Code*
- **NFPA 101** *Life Safety Code®*
- **NFPA 5000** *Building Construction and Safety Code*

KNOWLEDGE OBJECTIVES

After studying this chapter, you will be able to:

1. Describe an occupant load.
2. Describe issues with the means of egress at an existing occupancy, including locked doors, maintenance, and obstructions.
3. Describe how to ensure that emergency access for an existing site meets the policies of the AHJ, and how to report a deficiency of such.

SKILLS OBJECTIVES

After studying this chapter, you will be able to:

1. Observe, recognize, and report occupant load problems to the AHJ.
2. Identify the emergency access requirements and report deficiencies per the policies of the AHJ.
3. Observe and recognize problems, and make decisions related to means of egress.

You Are the First Responder Inspector

As a newly assigned first responder inspector, you are performing annual business inspections. Your assigned inspections bring you to a local tavern. You enter the tavern and notice the posted occupant load is 75 people. Quickly counting the crowd you notice the occupant load is considerably more. Further, you also notice that the side exit door is obstructed with tables.

1. What is the first step you should do regarding the overcrowding?
2. The occupancy has a total of three exits. What effect does one blocked exit have on the safety of the occupants?
3. Does the exit that is blocked from the inside impact emergency service access?

Introduction

While there is no one universal set of building, fire, or life safety codes in use across the United States, the National Fire Protection Association (NFPA), and the International Code Council (ICC) are the two most commonly used building and fire model codes. Each state either incorporates by reference a model building or fire code, or creates their own codes to meet their specific building and fire safety needs. Some cities, towns, and counties further modify the model codes to meet their particular needs. For example, model building and fire codes require fire sprinklers in most buildings at 12,000 square feet (1115 m^2). However, many communities across the county have modified that requirement require fire sprinklers at 5000 square feet (465 m^2) or even for every new building built, regardless of size. Some communities have required sprinklers in single family homes for over 20 years. However, it is only recently that the controversial subject of sprinklers in single family homes has made its way into the model codes.

NFPA 101, *Life Safety Code* is used in this text as the code of reference for determining the means of egress requirements. The *Life Safety Code* was first developed in 1913 by the Committee on Safety to Life. In 1927, it was named the *Building Exits Code*, and later named the *Life Safety Code*. It is neither a building nor fire code. There are some building and fire code requirements, but they are related to safely exiting a building, not the construction or protection of the building, as such. The *Life Safety Code* forms the basis for most of the egress requirements contained in the model building and fire codes. The *Life Safety Code* also addresses the construction, protection, and occupancy features necessary to minimize dangers to life from the effects of fire, including smoke, heat, and toxic gases. It also establishes minimum criteria for the design of egress facilities to allow for the prompt escape of occupants from buildings or to safe areas within buildings.

Occupant Load

The **occupant load** is the number of people who might occupy a given area. The occupant load reflects the maximum number of people anticipated to occupy the building space(s) at any given time. The occupant load must not be based only on normal occupancy, because the greatest hazard can occur when an unusually large crowd is present, which is a condition often difficult for the inspector to control.

While creating the occupant load figure is not too difficult for the fire inspector, the occupant load figure can be fluid. For example, a large open room will have

PRO TIPS

Navigating the *Life Safety Code* can be confusing at times, especially to new first responder inspectors. To assist, the *Life Safety Code* has an annex that contains explanatory material. This annex offers great insight to what the authors of the *Life Safety Code* intended. Areas in the *Life Safety Code* that have annex material are marked with an asterisk (*). By referring to this material found in the back of the *Life Safety Code*, you may be able to make more informed decisions. A solid bar in the margin indicates a change from the previous edition of the *Life Safety Code*. It will not tell you what the change was, only that a change occurred from one edition to another.

> **PRO TIPS**
>
> You must have access to and be fully knowledgeable of the codes, standards, and regulations in effect in your jurisdiction.

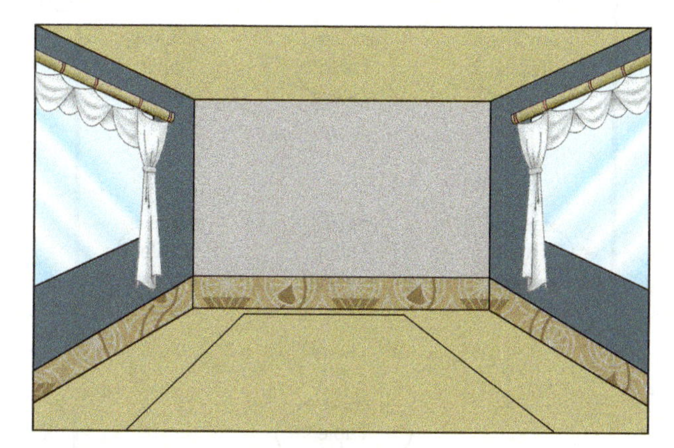

FIGURE 5-1 A large open room will have a larger occupant load then a room filled with tables and chairs.

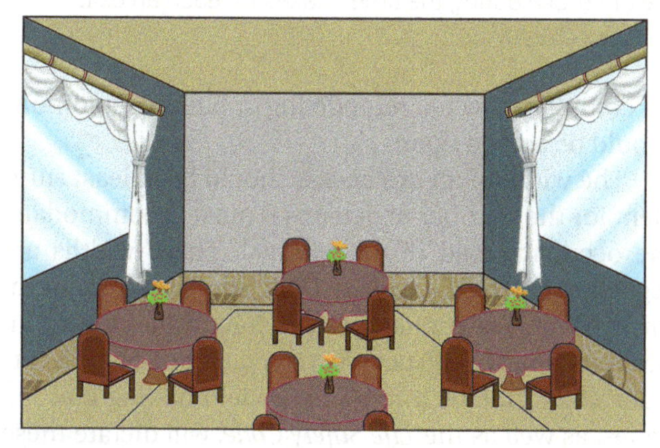

FIGURE 5-2 Chairs and tables take up floor space and reduce the occupant load in a room.

one occupant load (**FIGURE 5-1**). When it is filled with table and chairs, such as in a banquet, it has another occupant load figure (**FIGURE 5-2**). When posting occupant loads, some fire departments indicate two numbers. The first is for a table and chair configuration and the second for a concentrated area.

Ideally the design professional for the building should determine the occupant load figure based upon the uses within the building and the egress capacity of the designated exits, and include that on the building plans that are submitted to the authority having jurisdiction (AHJ). The first responder inspector will receive the occupant load, typically in the form of a permit, for inspection and verification purposes.

> **PRO TIPS**
>
> Occupant load is based on use of a space, not the occupancy. A school is an educational occupancy, but it has spaces of classrooms, kitchens, libraries, laboratories, swimming pools, etc. Occupant load is determined by the nature of the use of a building or space, the amount of space available for that use, and the egress capacity of the exits. Since different generic uses are characterized by different occupant densities, the *Life Safety Code* has established occupant load factors for each use.

> **PRO TIPS**
>
> Once occupant loads are posted, it is not uncommon that they are exceeded. Much of controlling this rests with the business manager or, in the case of larger public assembly situations, a crowd manger. It is difficult for a first responder inspector to routinely check the occupant load of a bar or nightclub at 1 AM. In locations where there are many late evening assembly occupancies or where there are documented instances of overcrowding, fire departments often will send first responder inspectors to conduct basic inspections during those peak hours to check for overcrowding or other fire violations.

Means of Egress

The *Life Safety Code* and the model building codes use the term **means of egress**. A means of egress is a continuous path of travel from any point in a building to a public way that is safely away from the building. A means of egress consists of three separate and distinct parts:

- **Exit access**—That portion of a means of egress that leads to the entrance of an exit. In other words, the exit access is the travel anywhere within the building to the exit from the building. This may include travel within corridors, on stairs, or traversing open floor areas. There are limits an occupant can travel within the exit access to actually reach an exit, which is called the "Maximum Travel Distance."

- **Exit**—That portion of a means of egress that is between the exit access and the exit discharge. An exit may be comprised of vertical and horizontal means of travel. For example, entering an exit stairway on the fourth floor would end the exit access for that floor and begin the exit even though there are many floors of stairs to walk until the exit discharge is reached.
- **Exit discharge**—That portion of a means of egress between the termination of the exit and a public way. In other words, the exit discharge is the area between the exit and the nearest public way. All exits must terminate at a public way. While the requirements for exit discharge are vaguely defined, the entire distance must be identifiable, reasonably direct, and essentially unimpeded. This could include such things as removing the accumulation of snow and ice during the winter, making certain the terrain is even, and removing obstacles that might hinder movement to the public way. Most designs will exit onto a walkway that leads to the public way.

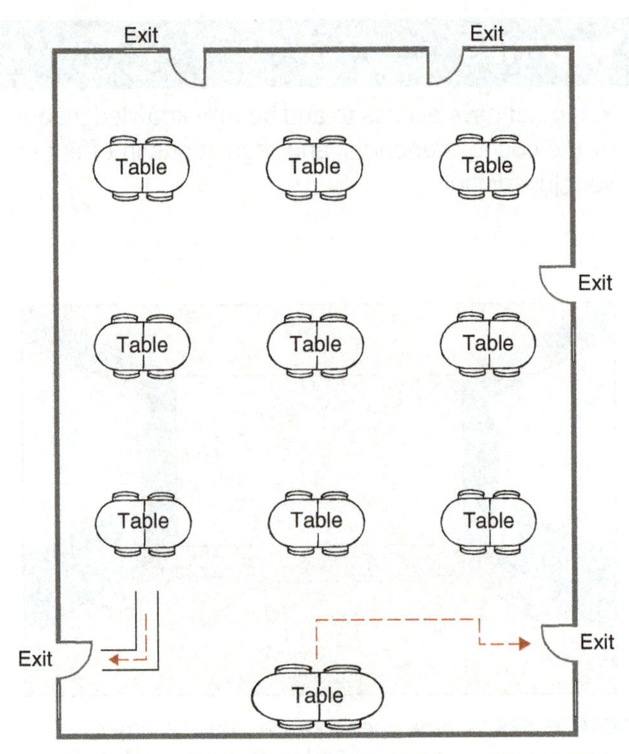

FIGURE 5-3 The length of the exit access establishes the travel distance to an exit, an extremely important feature of a means of egress, since an occupant might be exposed to fire or smoke during the time it takes to reach an exit.

Exit Access

The exit access may be a corridor, aisle, balcony, gallery, room, porch, or roof. Basically it is anything that takes you to an exit. The length of the exit access establishes the travel distance to an exit, an extremely important feature of a means of egress, since an occupant might be exposed to fire or smoke during the time it takes to reach an exit (**FIGURE 5-3**). The maximum travel distance to an exit is regulated by the *Life Safety Code* and is the distance an occupant is allowed to travel until an exit is encountered. The average maximum travel distance is 200 in. (61 m), but this distance varies with the occupancy and varies with the presence of fire suppression systems.

The travel distance must be measured from the most remote point in a room or floor area to an exit. In most cases, the travel distance can be increased up to 50 percent if the building is completely protected with an approved supervised automatic sprinkler system. This is something that must be noted during the plan review stage of construction. The need to shorten excessive travel distances must be made at this point because it could be almost impossible to make changes once the building is completed. If, in the course of a fire inspection in a completed building, the travel distance seems too far, it should be measured for compliance. If there is a problem, corrections should be made. Compliance may be just a matter of reconfiguring a furniture layout. If immediate corrections are not possible, then a plan for compliance when remodeling or building additions should be agreed upon.

The width of an exit access should be at least sufficient for the number of persons it must accommodate: the occupant load of the room(s). These widths are regulated by the *Life Safety Code* and have minimum dimensions. For example, if you have an area with an occupant load of 100 people, you could not put in a corridor of 20 in. (508 mm). The model building and fire codes, as well as the *Life Safety Code*, will dictate these minimum requirements.

In some occupancies, the width of the access is determined by the activity in the occupancy. One example is a new hospital, where patients may need to be moved in beds. The corridors in the patient areas of the hospital must be 8 ft (2.4 m) wide to allow for a bed to be wheeled out of a room and turned 90 degrees.

As these occupancies and widths have been determined for you by the *Life Safety Code*, it is only necessary that you understand that the dimensions change based on the occupancy and you must know where to look for the proper widths in the *Life Safety Code*. As these items relate to exiting, the specifics will be found in chapters dealing with means of egress requirements.

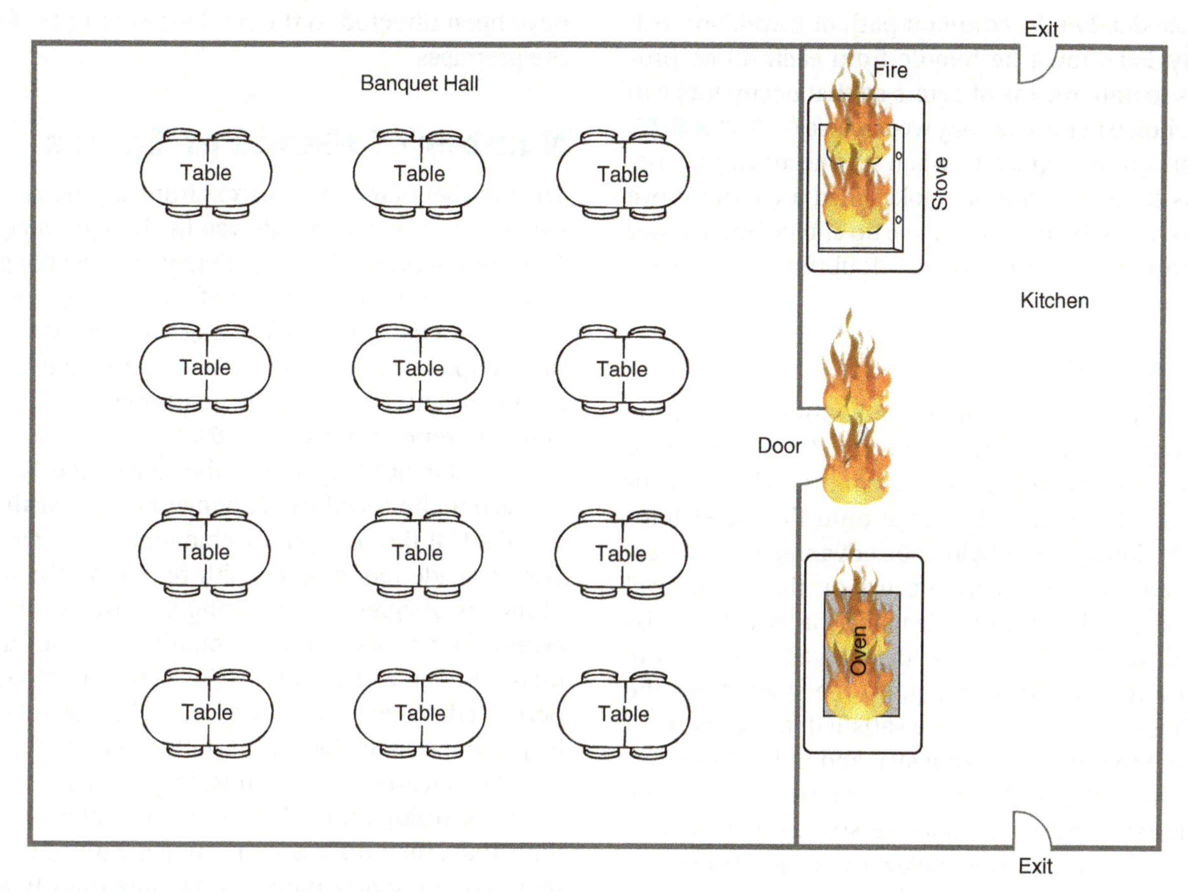

FIGURE 5-4 If the access passes through an area containing a fire hazard, the principles of free and unobstructed exit access are violated.

A fundamental principle of exit access is a free and unobstructed way to the exits. If the access passes through a room that can be locked or through an area containing a fire hazard more severe than is typical of the occupancy, the principles of free and unobstructed exit access are violated. A good example is a restaurant. If you are in the seating area and the main exit is through the front door, you would not be allowed to pass through the kitchen to get to the second exit. Obviously, a kitchen has a higher hazard than sitting at tables, and thus the exit would not be allowed for patrons (**FIGURE 5-4**). However, the kitchen exit is a viable exit for those working in the kitchen, as they are not moving through a higher hazard.

The floor of an exit access should be level. If this is not possible, differences in elevation may be overcome by a ramps or stairs. Where only one or two steps are necessary to overcome differences in level, in an exit access, a ramp may be preferred, because people may trip in a crowded corridor and fall on the stairs if they do not see the steps or notice that those in front of them have stepped up or down.

Exit

Examples of exits are doors leading directly outside at ground level, such as the front door of a business, or through a protected passageway to the outside at ground level. Examples of protected passageways include smokeproof towers, protected interior and outside stairs, exit passageways, enclosed ramps, and enclosed escalators or moving walkways in existing buildings. The actual exit may be as narrow as the width of a door, as in the case of a door from a retail area that leads directly to the outside, or it could be many, many floors, in the case of multi-floored buildings. In the case of multi-floored buildings, enclosed stairwells must meet certain requirements. As soon as you enter one of those stairwells, the exit access ends, you are in the exit stage, regardless of how many floors it may take to reach outside safety. Elevators are not accepted as exits, except in very special circumstances.

The specific placement of exits is a matter of design judgment, given the specifications of travel distance, allowable dead ends, common path of travel, and exit capacity. Exits must be remote from each other, providing separate means of egress so that occupants can have a choice in the exit they wish to use (**FIGURE 5-5**). This concept is important when it is necessary for occupants to leave a fire or smoke-filled area and move toward an exit. If occupants have no choice but to enter the fire area to reach an exit, it is doubtful whether they would be willing to do so.

Exit Discharge

Ideally, all exits in a building should discharge directly to the outside or through a fire-rated passageway to the outside of the building. A maximum of 50 percent of the exit stairs may discharge onto the street floor of the building. The obvious disadvantage of this arrangement is that if a fire occurs on the street level floor, it is possible for people using the exit stairs discharging to that floor to be discharged into the fire area. If any exits do discharge to the street floor the following conditions must be satisfied: the exits must discharge to a free and unobstructed public way outside of the building, the street floor must be protected by automatic sprinklers, and the street floor must be separated from any floors below by construction having a 2-hour fire resistance rating.

A free and unobstructed route to the public way is key. You cannot direct occupants to areas where they will be trapped by fencing. A common example of this would be an apartment complex that has a recreation building with an outside swimming pool. Many times the occupant load of the building will be greater than 50 thus requiring two exits. Often one exit may lead into the pool area. The difficulty with this arrangement is that because the owners do not want people entering the pool through a gate in the fence, none will be provided. This is unsafe because occupants who have been directed to the pool area cannot safely exit the premises.

FIGURE 5-5 Exits must be remote from each other, providing separate means of egress so that occupants can have a choice in the exit they wish to use.

Number of Means of Egress

The number of means of egress from any area is at least two, unless specifically allowed by the *Life Safety Code*. The first chapters of the *Life Safety Code* list the general requirements on the number of means of egress. Chapters in the *Life Safety Code* listing specific requirements for occupancies take precedence over general requirements. If nothing specific is mentioned, then the general requirement prevails (**TABLE 5-1**). This is why it is important to not only refer to the general requirements, but also to the specific occupancy, new or existing.

Most of the occupancy chapters in the *Life Safety Code* provide redundancy with respect to the number of means of egress by requiring at least two means of egress. Some occupancies identify specific arrangements under which only a single means of egress is permitted. Where large numbers of occupants are to be present on any floor or portion of a floor, then more than two means of egress must be provided.

If the occupant load is more than 500 but not more than 1000, then no less than three means of egress are required. If the occupant load is more than 1000, then no less than four means of egress are required. Several occupancies establish not only the minimum number of means of egress, but also the minimum number of actual exits that must be provided on each floor.

In most occupancies, meeting the requirements for egress capacities and travel distances means the required minimum number of means of egress will automatically be met. However, in occupancies characterized by high occupant loads, such as assembly and mercantile occupancies, or with unusual geometry that increases distance to exits, such as many

TABLE 5-1 Number of Exits per Occupancy

Occupant Load	Minimum Number Exits
0 to 499	2 (with limited exceptions for less than 50)
500 to 999	3
1000 or more	A minimum of 4

buildings in Las Vegas, compliance with requirements for more than two exits per floor might require specific attention.

Multi-Story Buildings

Similar to the procedure for verifying required egress capacity, the number of required exits is based on a floor-by-floor consideration, rather than the accumulation of the occupant loads of all floors. For example, if the fourth floor of a building has an occupant load of 700, it requires three exits. If the third floor of the same building has an occupant load of 400, it would require only two exits. Regardless of the fact that the two floors together have an occupant load in excess of 1000, four exits are not required. However, the number of exits cannot decrease as an occupant proceeds along the egress path. The three exits required from the fourth floor in this example cannot be merged into two exits on the third floor, even though the third floor requires only two exits. The number of required exits must be carried out down to the ground floor.

PRO TIPS

The first responder inspector's skill set involves recognizing existing egress and occupant load issues, and reporting deficiencies. The fire inspector's skill set involves calculating new egress requirements and occupant loads. Both the first responder inspector and fire inspector have the same goal: occupant safety.

PRO TIPS

You should have a working understanding of common terms and how to assess the correct distances. It would not be uncommon that during an inspection a building owner would ask your opinion about some remodeling being contemplated. Knowledge of the terms may raise red flags as the proposed plan is told to you and this could aid the owner's planning. A **dead end corridor** exists when an occupant enters a corridor and, finding no exit, is forced to retrace the path traveled to reach a choice of egress travel paths. Although relatively short dead ends are permitted by the *Life Safety Code*, it is better practice to eliminate them wherever possible, as they increase the danger of persons being trapped during a fire.

Common Path of Travel

A **common path of travel** is the distance an occupant must walk until there is a decision of which means of egress to use. Consider an office with one door opening into a corridor with exits at each end. The distance it takes to get through the office to the corridor is the common path of travel. Once the corridor is reached, a choice of what exit to use can be made.

Means of Egress Elements and Arrangements

Means of egress elements are the components of the means of egress including exit access, exit enclosures, exit discharges, stairways, ramps, doors, hardware, exit markings, illumination, etc.

Doors

All exits must have doors; however, not every door is an exit. When a door is marked as an exit, certain requirements in the *Life Safety Code* take effect such as identification, clear access, locking, widths, etc. Other doors, not classified as exits do not fall into these requirements. Take a small shopping strip with three small stores as an example. Each store has it front and rear exit. One business decides to take over the other two stores, and have one large, open store. The size of the store might still only require two remote exits. While individually each store required the rear exit, when combined, all three are not needed. However, if the doors continue to have exit signs, denoting them as exits, they must be maintained as such as people will be directed to those doors in an emergency. If they were to remove the exit signage from two of the doors, those two doors simply become convenience doors, and the *Life Safety Code* requirements for exit doors do not apply.

Doors should be side-hinged or pivoted swinging type and should swing in the direction of exit travel, except in small rooms when allowed by the *Life Safety Code* (**FIGURE 5-6**). Horizontal sliding, vertical, or rolling doors can be used as means of egress in certain occupancies. In assembly occupancies and in schools, panic hardware equipped with latches must be installed on all egress doors that serve rooms with occupant loads of 100 or more.

When doors protect exit facilities, as in stairway enclosures and horizontal exits, they normally must be kept closed to limit the spread of smoke. If open, they must be closed immediately in case of fire. Although

FIGURE 5-6 Doors should be side-hinged or pivoted swinging type and should swing in the direction of exit travel.

FIGURE 5-7 Panic hardware.

ordinary, fusible-link operated devices to close doors in case of fire are designed to close in time to stop the spread of fire, they do not operate fast enough to stop the spread of smoke and are therefore not permitted by the *Life Safety Code.*

Sometimes, people keep self-closing doors open with hooks or with wedges under the door. Doors also can be blocked open to provide ventilation, for the convenience of building maintenance personnel, or to avoid the accident hazard of swinging doors. The following measures have been provided in the *Life Safety Code* to alleviate these unallowable situations:

1. Doors that are normally kept open can be equipped with door closers and automatic hold-open devices that release the doors and allow them to close upon activation of the fire alarm system.
2. Doors that are normally closed can be equipped to open electrically or pneumatically when a person approaches the door, as long as precautions are used to prevent the door from automatically opening when there is smoke in the area.
3. Doors that normally are closed can be opened and held open manually by monitors, as in schools.
4. Smokeproof towers that protect against smoke can be used, even if the dooSrs are open.

In the event of electrical failure, the door must, however, close and remain closed unless it is opened manually for egress purposes.

Another major maintenance difficulty with exit doors are exterior doors that are locked to prevent unauthorized access. The *Life Safety Code* specifies that when the building is occupied, all doors must be kept unlocked from the side from which egress is made. If a door is locked it must be opened without the use of a key or any special knowledge, and with only one action. In other words, if a door is deadbolted closed, a person inside should be able to unlock the door by simply opening the handle. This setup is very common on hotel doors. Other measures to prevent unauthorized use of exit doors include:

- An automatic alarm that rings when the door is opened
- Visual supervision such as wired-glass panels, closed circuit television, and mirrors, which may be used where appropriate
- Automatic photographic devices to provide pictures of users

A single door in a doorway should not be less than 32 in. (813 mm) wide in new buildings and 28 in. (711 mm) in existing buildings. To prevent tripping, the floor on both sides of the door should have the same elevation for the full swing of the door.

Panic Hardware

Egress doors in assembly and educational occupancies, such as schools or movie theaters, must be equipped with panic hardware when the occupant load exceeds 100 (**FIGURE 5-7**). **Panic hardware** is defined as a door-latching assembly incorporating a device that releases the latch upon the application of force in the direction of egress travel. Panic hardware devices are designed to facilitate the release of the latching device on the door when a pressure of not more than 15 lb (6.8 kg) is applied in the direction of exit travel. Even though the door may be locked, pressing on this hardware will open the door. Panic hardware that has been tested and listed for use on fire-protection-rated

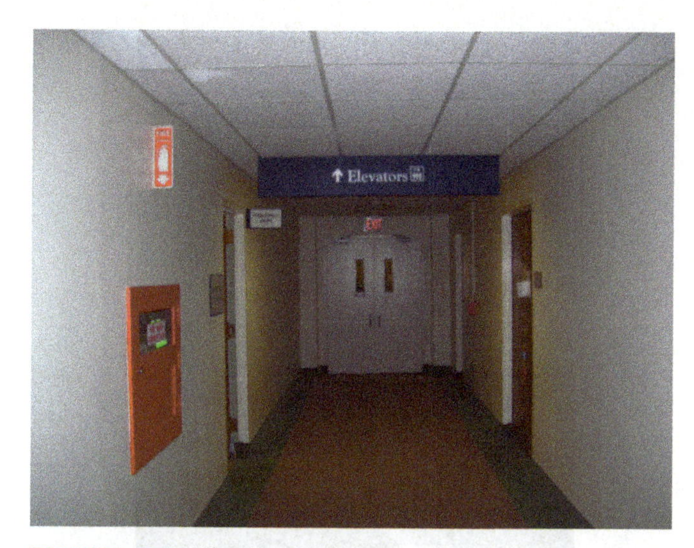

FIGURE 5-8 A horizontal exit.

doors is called "fire exit hardware." The panic hardware is the fire rated item. If panic hardware is needed on fire-protection-rated doors, only fire exit hardware can be used.

Horizontal Exits

A **horizontal exit** is a means of egress from one area to an area of refuge in another area on approximately the same level (**FIGURE 5-8**). Typically these means of egress are through 2-hour fire barriers. With a horizontal exit, space must be provided in the area of refuge for the people entering the refuge area in addition to the normal occupant load. The *Life Safety Code* recommends 3 square feet (0.28 m^2) of space per person, with the exception of healthcare and detention and correctional occupancies, where 6 to 30 square feet (0.56 to 2.79 m^2) of space is recommended. This is information that can be researched in the *Life Safety Code* should there be any question regarding how many people can occupy a horizontal exit.

Horizontal exits cannot comprise more than 50 percent of the total required exit capacity, except in healthcare facilities, where horizontal exits may comprise 66 percent of the total required exit capacity, or in detention and correctional facilities, where horizontal exits can comprise 100 percent of the total exit capacity. Horizontal exits are used in many healthcare facilities where the evacuation of patients over stairs is difficult, if not impossible, especially when considering patients in a hospital are confined to beds, many with ventilators or other machines. A horizontal exit arrangement within a single building and between two buildings is illustrated in **FIGURE 5-9**.

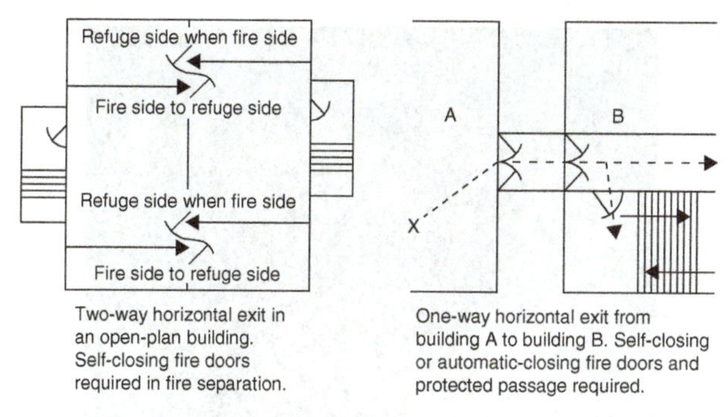

FIGURE 5-9 Types of horizontal exits.

Stairs

Exit stairs are designed to minimize the danger of falling, because a person falling on a stairway could result in the exit being blocked. Stair width must be based on calculated occupant loads. There should be no decrease in the width of the stairs along the path of travel, since this could create congestion or a bottle neck.

Steep stairs are dangerous and stair treads must be deep enough to give good footing. For new stairs, the *Life Safety Code* specifies a maximum 7 in. (178 mm) height (rise) and a minimum 11 in. (279 mm) tread (run) (7 to 11 stairs). Landings should be provided to break up any excessively long individual flight. Stairs can only be a maximum of 12 ft (366 cm) in length before a landing is required. Continuous railings are required for new stairs. These railings must also start before the stairs begin and terminate after the stairs end (**FIGURE 5-10**). New stairs more than 60 in. (1.5 m) wide need to have one or more center rails.

Stairs can serve as exit access, exit, or exit discharge. When used as an exit, they must be in an enclosure that meets exit enclosure requirements or be outside the building and properly protected. Outside stairs must comply with the requirements for exterior stairs and be arranged so that persons who fear heights will not be reluctant to use them. They should not be exposed to fire conditions originating in the building, and, where necessary, should be shielded from snow and ice. Exterior stairs should not be confused with fire escape stairs (**FIGURE 5-11**).

Stair enclosures involve the principles that are designed to limit fire and smoke spread. Doors or openings from each story are necessary to prevent the stairway from serving as a flue for products of combustion. In general, stairway enclosures should include not only the stairs, but also the path of travel from the bottom of the stairs to the exit discharge, so that occupants

FIGURE 5-10 Continuous railings are required for new stairs.

have a protected, enclosed passageway all the way to the public area. The stair enclosure should be of 1-hour construction when connecting three or fewer floors and of 2-hour construction when connecting four or more floors. Storage in egress stairways is prohibited by other sections of the code because it reduces the egress capacity and potentially provides a source of fuel.

Smokeproof Enclosure

Smokeproof enclosures are stair enclosures designed to limit the spread of the products of combustion from a fire. Smokeproof enclosures provide the highest protected type of stair enclosure recommended by the *Life Safety Code*. The stair tower is only accessed through balconies open to the outside air, vented vestibules, or mechanically pressurized vestibules, so that smoke, heat, and flame will not spread readily into the tower even if the doors are accidentally left open.

Ramps

Ramps, enclosed and otherwise, arranged like stairways, are sometimes used instead of stairways where there are large crowds and to provide both access and egress for nonambulatory persons. To be considered safe, exits ramps must have a very gradual slope. An example of a very common ramp would be at large scale professional sporting events, where many thousands of people are leaving the venue at one time (**FIGURE 5-12**).

A.

B.

FIGURE 5-11 Exterior stairs should not be confused with fire escape stairs. **A.** Exterior stairs. **B.** Fire escape stairs.

Exit Passageways

A hallway, corridor, passage, tunnel, or underfloor or overhead passageway may be used as an exit passageway, providing it is separated and arranged according to the requirements for exits. The use of a hallway or

FIGURE 5-12 Exit ramp.

corridor as an exit passageway requires close review. The *Life Safety Code* specifies that an exit enclosure should not be used for any purpose that could interfere with its value as an exit. For example, in an industrial occupancy, the use of a gasoline-powered forklift in a corridor designated as an exit passageway would violate the intent of the *Life Safety Code*. Also, penetration of the enclosure by ductwork and other utilities is typically not allowed by the *Life Safety Code*.

In addition, multiple doors in a corridor used as an exit increases the likelihood that a door could fail to close and latch, resulting in fire contamination. The door openings in exit enclosures should be limited to those necessary for access to the enclosure from normally occupied spaces. Therefore, doors and other openings to spaces such as boiler rooms, storage areas, trash rooms, and maintenance closets are not allowed in an exit passageway.

An exit passageway should not be confused with an exit access corridor. Exit access corridors do not have as stringent construction protection requirements as do exit passageways, since they provide access to an exit rather than being an actual component of the exit.

Fire Escape Stairs

Fire escapes should be stairs, not ladders. Fire escapes are, at best, a poor substitute for standard interior or exterior stairs. The *Life Safety Code* only permits existing fire escapes and only in existing buildings.

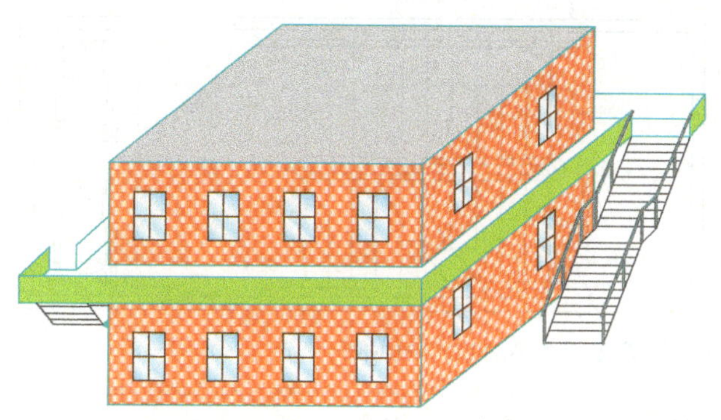

FIGURE 5-13 Fire escape stairs.

Fire escapes can create a severe fire exposure to people if flames in lower levels come through windows, blocking the path beneath them (**FIGURE 5-13**). The best location for fire escapes is on exterior masonry walls without exposing windows, with access to fire escape balconies by exterior fire doors. Where window openings expose fire escapes, shatter-resistant wired-glass in metal sashes should be used to provide maximum protection for the fire escape. When the building has an automatic sprinkler system the fire exposure hazard to personnel on fire escapes is minimized. An obvious problem with fire escape in northern climates is that outdoor fire escapes may be obstructed by snow and ice.

Escalators, Moving Walkways, and Elevators

Escalators are not recognized as an acceptable means-of-egress component in new construction. In some select cases they may be allowed in existing buildings, but the *Life Safety Code* should be referenced for specific details. Moving walkways also may be used as means of egress if they conform to the general requirements for ramps, if inclined, and for passageways, if level.

Elevators are not generally recognized as exits. However, elevator lobbies are permitted to be used if they have all of the requirements in place, to serve areas of refuge for the mobility impaired. Elevator cars may not be used. The *Life Safety Code* also recognizes elevators, under very limited conditions, as the second exit from limited access towers such as Federal Aviation Administration (FAA) control towers.

Areas of Refuge

An **area of refuge** is a protected area in a building designed for use by people with mobility impairments (**FIGURE 5-14**). It is a staging area that provides

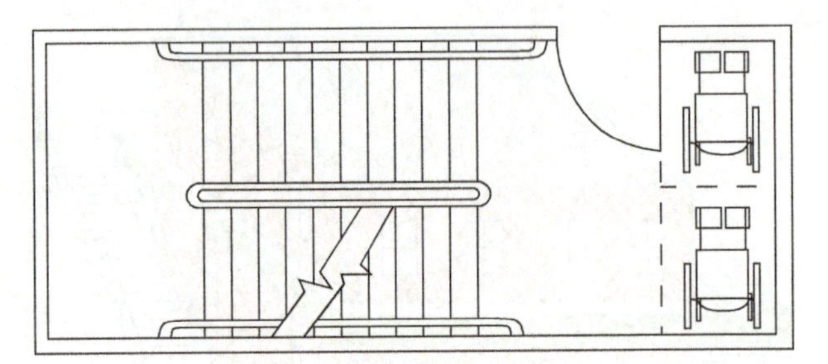

FIGURE 5-14 Area of refuge.

Reproduced with permission from NFPA's Fire Protection Handbook®, Copyright © 2008, National Fire Protection Association.

relative safety to its occupants until they can be completely evacuated. While not common, the *Life Safety Code* states that areas of refuge may be in a different building, accessed using corridors, balconies, etc. It further states that any floor with an approved, supervised sprinkler system is an area of refuge.

Ropes and Ladders

Ropes and ladders are not recognized as a substitute for standard exits from a building. There only possible use is in existing one- and two-family dwellings where it is impractical to add a secondary means of escape. In this case, a suitable rope or chain ladder, or a folding metal ladder may be suitable rather than having to leap from a window. However, it should be recognized that aged, infirm, very young, and physically handicapped persons cannot use ladders, and that if the ladder passes near or over a window in a lower floor, flames from the window can prevent the use of the ladder.

In the situations where these items are used, education of the public is key. It needs to be reinforced that the only time to leave through a window is as a last resort—when smoke or fire is in the room, posing an immediate threat. It is important to stress that a closed door, even a bedroom door, can keep fire out for about 10 minutes. Remember, just because you cannot exit down the hallway to get out, does not mean you need to leap out of a second floor window. Open the window, prepare to leave, but do so *only* when actually physically threatened by the fire.

Windows

Windows are not exits. They may be used as access to fire escapes in existing buildings if they meet certain criteria concerning the size of the window opening and the height of the sill from the floor. Windows may also be considered a means of escape from certain residential occupancies, if they meet minimum requirements specified in the *Life Safety Code*.

FIGURE 5-15 Exit lighting.

© Catalin D/Shutterstock

Windows are required in school rooms subject to student occupancy, unless the building is equipped with a standard automatic sprinkler system, and in bedrooms in one- and two-family dwellings that do not have two separate means of escape. These windows are for rescue and ventilation and must meet the criteria for size of opening, method of operation, and height from the floor.

Exit Lighting

In buildings where artificial lighting is provided for normal use, the illumination of the means of egress is required to ensure that occupants can see to evacuate the building quickly (**FIGURE 5-15**). The intensity of the illumination should be not less than 1 footcandle (10.77 lu/m^2) measured at the floor. It is desirable that such floor illumination be provided by lights recessed in the wall and located approximately 1 ft (30.5 cm) above the floor because such lights are then unlikely to be obscured by the smoke that might occur during a fire. In auditoriums and other places of public assembly where movies or other projections are shown, the *Life Safety Code* permits a reduction in this illumination for the period of the projection.

Emergency Lighting

The *Life Safety Code* requires emergency power for illuminating the means of egress in most occupancies. Some examples of where emergency lighting is required are assembly occupancies; most educational buildings; healthcare facilities; most hotels and apartment buildings; Class A and B mercantiles; and business buildings based on occupant load and number of stories.

Well-designed emergency lighting using a source of power independent from that of the normal building power automatically provides the necessary illumination in the event of an interruption of power to normal lighting. The failure of the public power supply or the opening of a circuit breaker or fuse should result in the automatic operation of the emergency lighting system.

Reliability of the exit illumination is important. NFPA 70, the *National Electrical Code®*, details requirements for the installation of emergency lighting equipment. Battery-operated electric lights and portable lights normally are not used for primary exit illumination, but they may be used as an emergency source under the restrictions imposed by the *Life Safety Code.* Luminescent, fluorescent, or other reflective materials are not a substitute for required illumination, since they are not normally sufficiently intense to justify recognition as exit floor illumination.

Where electric battery-operated emergency lights are used, suitable facilities are needed to keep the batteries properly charged. Automobile-type lead storage batteries are not suitable because of their relatively short life when not subject to frequent recharge. Likewise, dry batteries have a limited life, and there is a danger that they may not be replaced when they have deteriorated. If normal building lighting fails, well-arranged emergency lighting provides necessary floor illumination automatically, with no appreciable interruption of illumination during the changeover.

The emergency lighting is not designed to take the place of regular building lighting, but to provide a means to illuminate the means of egress. Where a generator is provided, a delay of up to 10 seconds is considered tolerable. In the cases of a generator being used as emergency power, it typically will power the normal building lighting. In that case, separate emergency lighting packs are not needed. The normal requirement is to provide such emergency lighting for a minimum period of 1½ hours. In the case of a long power failure, it is common to receive calls from residents in apartment buildings stating the hallways are black. As mentioned above, the lighting lasts only for 90 minutes so people have time to safely exit the building. They are not designed for the duration of a power failure.

Exit Signs

All required exits and access ways must be identified by readily visible signs when the exit or the way to reach it is not immediately visible to the occupants (**FIGURE 5-16**). Directional "EXIT" signs are required in locations where the direction of travel to the nearest exit is not immediately apparent. The character of the occupancy will determine the actual need for such signs. In assembly occupancies, hotels, department stores, and other buildings with transient populations, the need for signs will be greater than in a building with permanent or semi- permanent populations because those permanent or semi-permanent populations are more familiar with the various exit routes. Even in permanent residential-occupancy buildings, signs are needed to identify exit facilities, such as stairs, that are not used regularly during the normal occupancy of the building. It is just as important that doors, passageways, or stairs that are not exits but are so located that they may be mistaken for exits be identified by signs with the words "NO EXIT," with "no" larger than "exit."

FIGURE 5-16 Exit sign.

Exit signs should be readily visible: prominent, well-located, and unobscured. Decorations, furnishings, or other building equipment should not obscure the visibility of these signs. The *Life Safety Code* does not make any specific requirement for sign color but requires that signs be of a distinctive color. The *Life Safety Code* specifies the size of the sign, the dimensions of the letters, and the levels of illumination for both externally and internally illuminated signs. The *Life Safety Code* also requires in specific occupancies that exit signs are needed at the on the wall close to the floor.

Maintenance of the Means of Egress

The provision of a standard means of egress with adequate capacity does not guarantee the safety of the occupants in the event of an evacuation if the means of egress are not properly maintained. Many building owners and property managers do not assign someone to be responsible for the maintenance of the means of egress. As a result you may find stairways used as storage for materials during peak sales or manufacturing periods. In apartment buildings, rubbish, bicycles, baby carriages, and other obstructions are often found in stairway enclosures. Exit doors may be found locked or with hardware in need of repair. Supplemental locking devices for intruder deterrence can also create safety hazards. NFPA 101 provides guidance for educational occupancies, requiring AHJ approval. Doors blocked open or removed from openings into stairway enclosures may permit rapid spread of smoke or hot gases throughout the building. Loose handrails and loose or slippery stair treads may cause persons evacuating a building to fall in the path of others seeking escape. Maintaining the means of egress in safe operating condition at all times is as important as the proper design of the egress system itself. The first responder inspector is primarily interested in the proper maintenance of the means of egress and its components.

WRAP-UP

CHAPTER SUMMARY

- The National Fire Protection Association (NFPA) and the International Code Council (ICC) are the two most commonly used building and fire model codes.
- Each state either adopts a model building or fire code, or creates their own codes to meet their specific building and fire safety needs.
- The occupant load reflects the maximum number of people anticipated to occupy the building space(s) at any given time.
- The occupant load figure can be fluid. A large open room will have one occupant load. When it is filled with table and chairs, as in a banquet hall, it has smaller occupant load figure. When posting occupant loads, some fire inspection units indicate the two potential occupant load figures.
- A means of egress consists of three separate parts: exit access, exit, and exit discharge.
- The exit access may be a corridor, aisle, balcony, gallery, room, porch, or roof. The length of the exit access establishes the travel distance to an exit, an extremely important feature of a means of egress, since an occupant might be exposed to fire or smoke during the time it takes to reach an exit.
- Examples of exits are doors leading directly outside at ground level, such as the front door of a business, or through a protected passageway to the outside at ground level. The latter includes smokeproof towers, protected interior and outside stairs, exit passageways, enclosed ramps, and enclosed escalators or moving walkways in existing buildings.
- Ideally, all exits in a building should discharge directly to the outside or through a fire-rated passageway to the outside of the building.
- At least two means of egress must exist in any area, unless specifically allowed by the *Life Safety Code*. The first chapters of the *Life Safety Code* list the general requirements on the number of means of egress.
- Means-of-egress elements are the components of the means of egress including exit access, exit enclosures, exit discharges, stairways, ramps, doors, hardware, exit markings, illumination, etc.
- In buildings where artificial lighting is provided for normal use, illumination of the means of egress is required to ensure that occupants can see to evacuate the building quickly.

KEY TERMS

Area of refuge An area that is either (1) a story in a building where the building is protected throughout by an approved, supervised automatic-sprinkler system and has not less than two accessible rooms or spaces separated from each other by smoke-resisting partitions; or (2) a space located in a path of travel leading to a public way that is protected from the effects of fire, either by means of separation from other spaces in the same building or by virtue of location, thereby permitting a delay in egress travel from any level. (NFPA 101)

Common path of travel The portion of exit access that must be traversed before two separate and distinct paths of travel to two exits are available. (NFPA 101)

Dead end corridor A passageway from which there is only one means of egress. (NFPA 301)

Exit That portion of a means of egress that is separated from all other spaces of a building or structure by construction or equipment as required to provide a protected way of travel to the exit discharge. (NFPA 101)

Exit access That portion of a means of egress that leads to an exit. (NFPA 101)

Exit discharge That portion of a means of egress between the termination of an exit and a public way. (NFPA 101)

Horizontal exit An exit between adjacent areas on the same deck that passes through an A-60 Class boundary that is contiguous from side shell to side shell or to other A-60 Class boundaries. (NFPA 301)

Means of egress A continuous and unobstructed way of exit travel from any point in a building or structure to a public way, consisting of three separate and distinct parts: (a) the exit access, (b) the exit, and (c) the exit discharge. A means of egress comprises the vertical and horizontal travel and includes intervening room spaces, doorways, hallways, corridors, passageways, balconies, ramps, stairs, enclosures, lobbies, escalators, horizontal exits, courts, and yards. (NFPA 101)

Occupant load The number of people who might occupy a given area.

Panic hardware A door-latching assembly incorporating a device that releases the latch upon the application of a force in the direction of egress travel. (NFPA 101B)

Smokeproof enclosures A stair enclosure designed to limit the movement of products of combustion produced by a fire. (NFPA 101)

You Are the First Responder Inspector

A local business on the ground floor of a five-story building; that uses large ovens for curing ceramics; was found to be holding open the doors into the exit stairway due to heat from the occupancy. They are using wooden wedges, as well as a fire extinguisher.

1. **What is the concern in a door opening into the egress stairway to vent hot air?**
2. **What could happen to the conditions within the exit stairway if a fire occurred within the business?**
3. **Is there a concern with using unlisted methods to hold open doors?**

CHAPTER

6

Fire Alarm and Detection Systems

NFPA 1030 THAT INCLUDES CHAPTER 6: FIRST RESPONDER INSPECTOR (NFPA 1031)

First Responder Inspector

- 6.5.7

ADDITIONAL NFPA STANDARDS

- **NFPA 72**, *National Fire Alarm and Signaling Code*

KNOWLEDGE OBJECTIVES

After studying this chapter, you will be able to:

1. Describe the basic components and functions of a fire alarm system.
2. Describe the basic types of fire alarm initiation devices and indicate where each type is most suitable.
3. Describe the first responder inspector's role in inspection fire detection systems.

SKILLS OBJECTIVES

After studying this chapter, you will be able to:

1. Determine the operational readiness of an existing fire alarm system.

You Are the First Responder Inspector

As a first responder inspector, you are conducting a routine inspection at a local bank. You notice a fire alarm panel beeping and the system malfunctioning. Turning to the manager walking with you, you ask if they know what's wrong. They reply that it's been beeping for a long time and they usually just press the silence button. Concerned, you request the most recent inspection report to verify operational status. They hesitate, admitting they don't remember the last inspection and wouldn't be able to find the report, raising serious safety concerns.

1. How should you remedy this situation?
2. How can you prevent this situation in the future?

Introduction

Fire alarm systems are critical for communicating vital information to the occupants and the outside world. Fire alarm systems can protect the occupants and the structure by providing early warning to both the occupants and firefighters which can minimize both loss of life and property damage during fires. This chapter will provide an overview of the types and features of fire alarm systems. You will also learn how devices may lend to impairment of the system when not properly maintained.

Fire Alarm and Detection Systems

Many new construction projects, including single family homes, require some sort of fire alarm and detection system. Fire alarm and detection components are integrated in a single system that can perform a myriad of different functions, using multiple devices. A fire detection system recognizes when a fire is occurring and activates the fire alarm system, which then alerts the building occupants and activates building functions such as recalling elevators, shutting down HVAC units, unlocking doors, releasing door hold-open devices, and, in most cases, notifying the fire department. Some fire detection systems also automatically activate fire suppression systems to control the fire.

Fire alarm and detection systems range from simple, such as a small one-zone system in a coffee shop, to complex fire detection and control systems for high-rise buildings (**FIGURE 6-1**). Many fire alarm and detection systems in large buildings also control other systems to help protect occupants and control the spread of fire and smoke. Although these systems can be complex, they generally include the same basic components.

FIGURE 6-1 Fire alarm and detection systems range from simple one zoned system to complex fire detection and control systems. As a first responder inspector, you need to understand the components for all types of systems.

Fire Alarm and Detection System Components

A fire alarm and detection system has three basic components: an alarm initiation device, an alarm notification device, and a control panel. The **alarm initiation device** is either an automatic or manually operated device that, when activated, causes the system to indicate an alarm. The **alarm notification appliance** is generally an audible device, more recently accompanied by a visual indication, that alerts the building occupants once the system is activated. The control panel links the initiation device to the notification appliance and performs other essential functions as required.

PRO TIPS

It is also very helpful to have an alarm matrix such the one listed in NFPA 72, *National Fire Alarm and Signaling Code*. An **alarm matrix** is a chart showing what will happen with the fire alarm and detection system when an initiating device is activated. Here, all the various types of devices are listed, including the inputs to the control panel and the possible outputs (functions) the panel can perform.

PRO TIPS

While the smoke alarms and smoke detectors seem to be one in the same there are slight differences between the terms. Smoke detectors are devices that are placed in a building and connected to a control panel. Smoke alarms refer to the single station units found in almost all homes.

Fire Alarm System Control Panels

Most fire alarm and detection systems have several alarm initiation devices in different areas and use both audible and visible devices to notify the occupants of an alarm. The **fire alarm control panel** serves as the "brain" of the system, linking the initiating devices to the notification devices.

The control panel manages and monitors the proper operation of the system. It can indicate the source of an alarm, so that responding fire personnel will know what activated the alarm and where the initial activation occurred. The control panel also manages the primary power supply and provides a backup power supply for the system. It may perform additional functions, such as notifying the fire department or central station monitoring company when the alarm system is activated, and may interface with other systems and facilities such as recalling elevators, unlocking stairwell doors to allow people in the stairs to re-enter the floors, releasing doors held open by magnetic hold open devices, and shutting down ventilation systems.

Control panels vary greatly depending on the age of the system and the manufacturer. For example, an older system may simply indicate that an alarm has

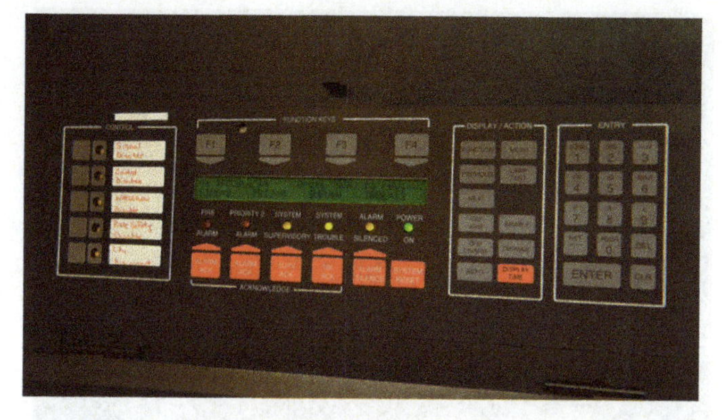

FIGURE 6-2 Most modern fire alarm control panels indicate the zone where an alarm was initiated.

been activated, whereas a newer system may indicate that alarm occurred in a specific zone within the building (**FIGURE 6-2**). The most modern panels actually specify the exact location of the activated initiation device. These panels are known as addressable panels.

Fire alarm control panels are used to silence the alarm and reset the system. These panels should always be locked, or in a room that is locked from the public. Many newer systems require the use of a password or key before the alarm can be silenced or reset. A problem with having a password is that the firefighters will need to know the passwords to potentially hundreds of alarm systems. It is much easier to have the control panel access gained through the use of a key, which can be kept in the key box at the building. Alarms should not be reset until the activation source has been found and checked by firefighters to ensure that the situation is under control. If the system is reset prior to identifying the problem, firefighters have no good way to determine the problem, and the activation may reoccur.

Some types of alarms require the activated devices to be reset prior to the control panel being reset. An example would be for activated manual pull stations. Other alarms, such as trouble or supervisory alarms, will reset the control panel automatically after the problem is resolved. A common example of this would be a power failure.

Many buildings have an additional display panel in a separate location, usually near the front door of the building. This panel, which is called a **remote annunciator**, enables firefighters to ascertain the type and location of the activated alarm device as they enter the building eliminating the need for firefighters to hunt down the control panel to determine the problem (**FIGURE 6-3**). There are two types of annunciators. The older style simply indicates the location of the

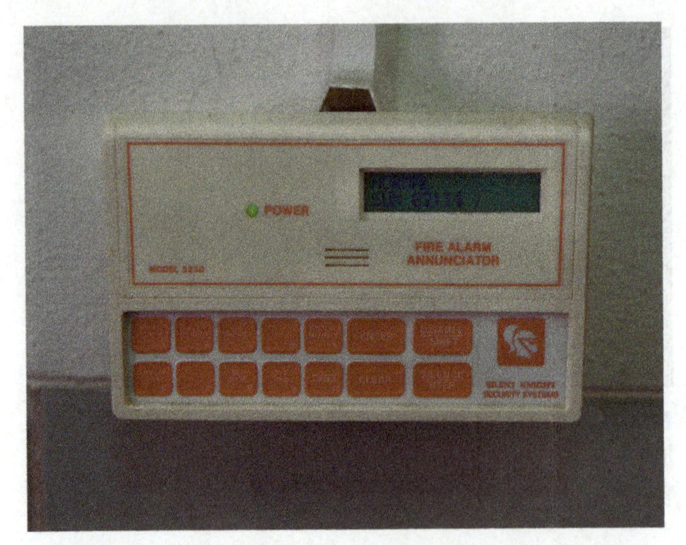

FIGURE 6-3 A remote annunciator allows firefighters to quickly determine the type and location of the activated alarm device.

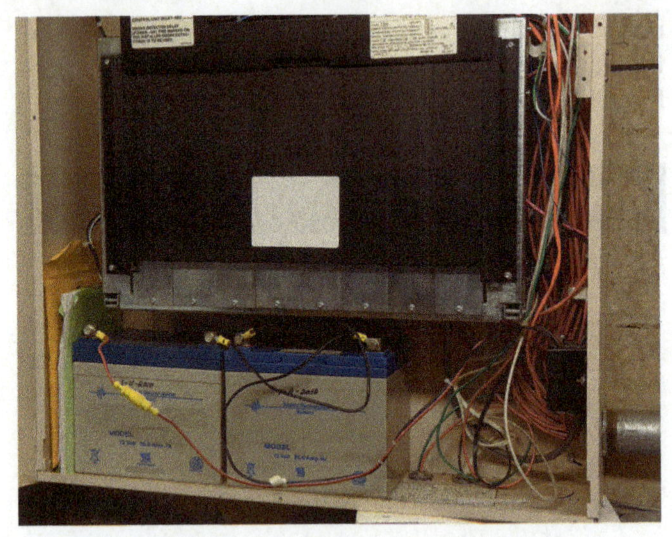

FIGURE 6-4 Model building and fire codes and their referenced standards require that alarm systems have a backup power supply, which is activated automatically when the normal electrical power is interrupted.

alarm. The newer annunciators are known as functional annunciators. This type of annunciator will provide as much detail about the problem as the control panel. In addition, it has the capability of acknowledging, silencing, and resetting the alarm. This allows the main control panel to be put in any location of the building, as full control of the alarm system is performed at the functional annunciator. These typically are found just inside the main entrance of the building.

The fire alarm control panel should also monitor the condition of the entire alarm system to detect any faults. Faults within the system are indicated by a trouble alarm, which shows that a component of the system is not operating properly and requires attention. It also monitors the integrity of the control panel's circuits. Trouble alarms do not activate the building's fire alarm but will make an audible sound and illuminate a light at the alarm control panel. They will also transmit a notification to a remote service location, such as a central station monitoring company.

Another type of alarm that the control panel will monitor is a supervisory alarm. Supervisory alarms sound when something changes from the normal ready condition within the system. For example, tamper switches are put on control valves of the fire sprinkler system, and when a control valve is closed—shutting down the sprinkler system—an alarm is sounded at the control panel and transmitted to the monitoring location. As with trouble alarms, this activation will not sound the general building's alarms.

A fire alarm system is usually powered by 110-volts, even though the system's appliances may use a lower voltage. Some extremely old alarm systems, however, require 110 volts for all components.

In addition to the normal power supply, the codes and standards require a backup power supply for all alarm systems (**FIGURE 6-4**). In most systems, a battery in the fire alarm control panel is activated automatically when the external power is interrupted. The model building and fire codes and their referenced standards will specify how long the system must be able to function on the battery backup. Typically 24 or 60 hours of backup power is provided. Many agencies use the 60-hour systems, which would cover the power loss in a building for an entire weekend. In large buildings, the backup power supply could be an emergency generator. If either the main power supply or the battery backup power fails, the trouble alarm should sound.

Depending on the building's size and floor plan, an activated alarm may sound throughout the building or only in particular areas. In high-rise buildings, fire alarm systems are often programmed to alert only the occupants on the same floor as the activated alarm as well as those on the floors immediately above and below the affected floor. If an initiating device is activated on any other floor, the alarms in that those areas would then activate as well. Some systems have a public address feature, enabling the fire alarm panel to play a prerecorded message, or for a fire department officer to provide specific instructions or information for occupants.

Residential Smoke Alarms

Current building and fire prevention codes require the installation of fire detectors with an alarm component in all residential dwelling units. **Single-station smoke alarms** are most commonly used (**FIGURE 6-5**). Millions of single-station smoke alarms have been installed in private dwellings and apartments, and countless numbers of lives have been saved due to them.

Smoke alarms can be either battery-powered or hard-wired to a 110-volt electrical system. Most building codes require hardwired, AC-powered smoke alarms with battery backup in all newly constructed dwellings; battery-powered units are popular for existing residencies. The major concern with a battery-powered smoke alarm is ensuring that the battery is replaced on a regular basis. The International Association of Fire Chiefs (IAFC) has a campaign to get home owners to change the batteries in their smoke alarms. The Change Your Clock, Change Your Battery campaign is designed to remind people to replace the batteries in the spring and fall when daylight saving time occurs.

Newer battery-powered smoke alarms are available with lithium batteries that will last for 10 years. These units have the battery sealed inside the detector to prevent it from being used somewhere else. As smoke alarms should be replaced after 10 years, this poses no problem.

The original codes stated that the smoke alarm should be placed in the hallway outside of the sleeping rooms. That concept would work well if bedroom doors were left open and smoke was allowed to work into the hallway and activate the smoke alarm. Since bedroom doors are closed a good portion of the time, the most up-to-date codes also require new homes to have a smoke alarm in every bedroom and on every floor level. They must be also interconnected so that, if the basement smoke alarm is activated, alarms will sound on all levels of the home to alert the occupants. This arrangement is called multiple-station smoke alarms.

FIGURE 6-5 The most common residential fire alarm system today is a single-station smoke alarm.

Many homes have added smoke and heat detectors as part of their home security systems. These devices operate just like the devices on an approved fire alarm system, except they are connected to a burglar alarm panel. These systems will require a passcode to set or reset the system and may be monitored by a central station. As home alarms are generally a burglar alarm system, it is unlikely that the fire department would have a password or code to reset the system.

Ionization versus Photoelectric Smoke Detectors

Two types of fire detection devices may be used in a smoke alarm to detect combustion:

- Ionization detectors are triggered by the invisible products of combustion.
- Photoelectric detectors are triggered by the visible products of combustion.

Ionization smoke detectors work on the principle that burning materials release many different products of combustion, including electrically charged microscopic particles. An ionization detector senses the presence of these invisible charged particles (ions).

An ionization smoke detector contains a very small amount of radioactive material inside its inner chamber. The radioactive material releases charged particles into the chamber, and a small electric current flows between two plates (**FIGURE 6-6**). When smoke particles enter the chamber, they neutralize the charged particles and interrupt the current flow. The detector then senses this interruption and activates the alarm.

Photoelectric smoke detectors use a light beam and a photocell to detect larger visible particles of smoke (**FIGURE 6-7**). They operate by reflecting the light beam either away from or onto the photocell, depending on the design of the device. When visible particles of smoke pass through the light beam, they interfere with the light beam, thereby activating the alarm.

Ionization smoke detectors are more common and less expensive than photoelectric smoke detectors. They react more quickly than photoelectric smoke detectors to fast-burning fires, such as a fire in

FIGURE 6-6 An ionization smoke detector.

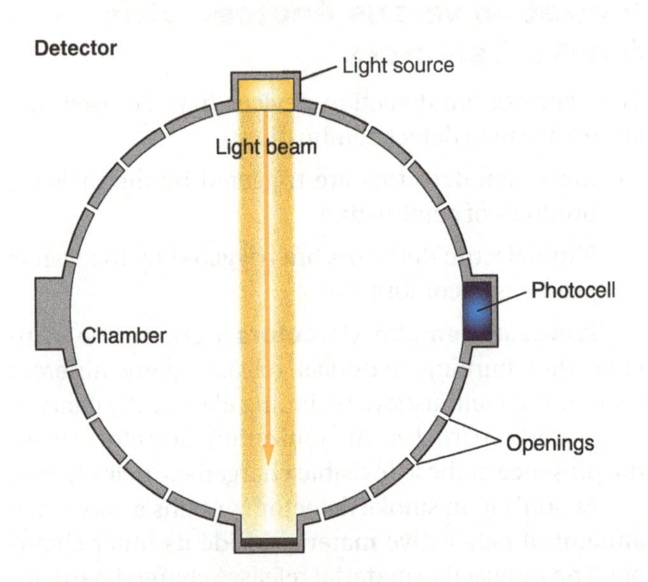

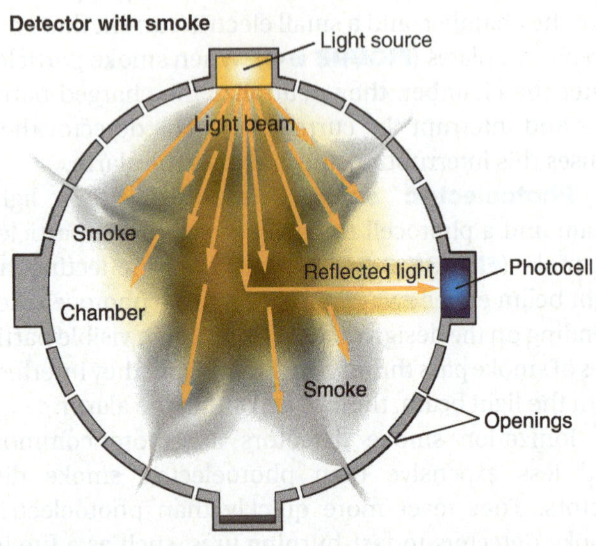

FIGURE 6-7 A photoelectric smoke detector.

FIGURE 6-8 Ionization smoke detectors react more quickly than photoelectric smoke detectors to a fast-burning fire, such as a fire in a wastepaper basket, which may produce little visible smoke.

a wastepaper basket, which may produce little visible smoke (**FIGURE 6-8**). On the downside, fumes and steam from common activities such as cooking and showering may trigger unwanted alarms.

Photoelectric smoke detectors are more responsive to slow-burning or smoldering fires, such as a fire caused by a cigarette caught in a couch, which usually produces a large quantity of visible smoke. They are less prone to false alarms from regular cooking fumes and shower steam than are ionization smoke detectors. Most photoelectric smoke alarms require more current to operate, however, so they are typically connected to a 110-volt power source.

Both ionization and photoelectric smoke detectors are acceptable life-safety devices. Indeed, in most cases, they are considered interchangeable. Some fire codes require that photoelectric smoke detectors be used within a certain distance of cooking appliances or bathrooms, to help prevent unwanted alarms.

Combination ionization/photoelectric smoke alarms are also available. These alarms will quickly react to both fast-burning and smoldering fires. They are not suitable for use near kitchens or bathrooms, because they are prone to the same nuisance alarms as regular ionization smoke detectors. **TABLE 6-1** indicates the general features and recommended applications for each type of smoke detector.

TABLE 6-1 Types of Smoke Detectors

Type of Detector	Features	Application
Ionization	Uses radioactive material within the device to detect invisible products of combustion.	Used to detect fires that do not produce large quantities of smoke in their early states. React quickly to fast-burning fires. Inappropriate for use near cooking appliances or showers.
Photoelectric	Uses a light beam to detect the presence of visible particles of smoke.	Used to detect fires that produce visible smoke. React quickly to slow-burning, smoldering fires.

Most ionization and photoelectric smoke detectors look very similar to each other. The only sure way to identify the type of alarm is to read the label, which is often found on the back of the case. An ionization alarm must have a label or engraving stating that it contains radioactive material.

From a first responder inspector's viewpoint, the best option is to recommend to home owners to install an alarm that has both ionization and photoelectric sensors. It should also be noted that the most common type of alarm, the ionization, has performed well for many years and has saved many lives. Any smoke alarm is better than no smoke alarm. As a result of the press that some studies have evoked, some jurisdictions have made a requirement that only photoelectric alarms are allowed to be used in homes. Always state your local jurisdiction's policies when making recommendations.

Alarm Initiation Devices

Alarm initiation devices begin the fire alarm process either manually or automatically. Manual initiation devices require human action; automatic devices function without human intervention. Manual fire alarm boxes are the most common type of alarm initiation devices that require human action. Automatic initiation devices include various types of heat and smoke detectors and other devices that automatically recognize the evidence of a fire.

Manual Initiation Devices

Manual initiation devices are designed so that building occupants can activate the fire alarm system on their own if they discover a fire in the building. Many older alarm systems could be activated only manually. The primary manual initiation device is the manual fire alarm box, or **manual pull-station** (**FIGURE 6-9**). Such a station has a switch that either opens or closes an electrical circuit to activate the alarm.

A.

B.

FIGURE 6-9 Several types of manual fire alarm boxes (also known as manual pull-stations) are available.

Pull-stations come in a variety of sizes and designs, depending on the manufacturer. They can be either single-action or double-action devices. **Single-action pull-stations** require a person to pull down a lever, toggle, or handle to activate the alarm. The alarm sounds as soon as the pull-station is activated. **Double-action pull-stations** require a person to perform two steps before the alarm will activate. They are designed to reduce the number of false alarms caused by accidental or intentional pulling of the alarm. The person must move a flap, lift a cover, or break a piece of glass to reach the alarm activation device. Designs that use glass are no longer in favor, because the glass must be replaced each time the alarm is activated and because the broken glass poses a risk of injury.

Once activated, a manual pull-station should stay in the "activated" position until it is reset. This enables responding firefighters to determine which pull-station initiated the alarm. Resetting the pull-station requires a special key, a small screwdriver, or an Allen wrench. The pull-station must be reset before the building alarm can be reset at the alarm control panel. A keyed pull station may be preferred as it can be kept in the building's locked key box. Small screwdrivers or Allen wrenches tend to get lost easily.

A variation on the double-action pull-station, designed to prevent malicious false alarms, is a single station pull box covered with a piece of clear plastic (**FIGURE 6-10**). These covers are often used in areas where malicious false alarms occur frequently, such as high schools and college dormitories. The plastic cover must be opened before the pull-station can be activated. Lifting the cover triggers a loud tamper alarm at that specific location, but does not activate the fire alarm system. Snapping the cover back into place resets the tamper alarm. The intent is that a person planning to initiate a false alarm will drop the cover and run when the tamper alarm sounds. In most cases, the pull-station tamper alarm is not connected to the fire alarm system.

FIGURE 6-10 A variation on the double-action pull-station, designed to prevent malicious false alarms, has a clear plastic cover and a separate tamper alarm.

Even automatic fire alarm systems can be manually activated by pressing a button or flipping a switch on the alarm system control panel. This is more useful for testing the system, as it would be impractical to expect someone to go to the control panel and activate the alarm manually during a fire.

Automatic Initiation Devices

Automatic initiation devices are designed to function without human intervention and will activate the alarm system when they detect evidence of smoke or fire. These systems can be programmed to transmit the alarm to the fire department or an on- or off-site monitoring facility, even if the building is unoccupied, and to perform other functions when a detector is activated.

Automatic initiation devices can use any of several types of detectors. Detectors are activated by smoke, heat, the light produced by actual flame, or specific gases created by fires.

Smoke Detectors. A **smoke detector** is designed to sense the presence of smoke and refers to a sensing device that is part of a fire alarm system. This type of device is commonly found in school, hospital, business, and commercial occupancies that are equipped with fire alarm systems (**FIGURE 6-11**).

FIGURE 6-11 Commercial ionization smoke detector.

TROUBLESHOOTING SMOKE ALARMS

Most problems with smoke alarms are caused by a lack of power, dirt in the sensing chamber, or defective alarms.

Power Problems

Smoke alarms require power from a battery or from a hardwired 110-volt power source. Some smoke alarms that are hard-wired to a 110-volt power source are equipped with a battery that serves as a backup power source.

The biggest problem with battery-powered alarms is a dead or missing battery. Some people do not change batteries when recommended, resulting in inoperable smoke alarms. People may also remove smoke alarm batteries to use the battery for another purpose or to prevent nuisance alarm activations. The solution in this situation is to install a new battery in the alarm.

Hard-wired smoke detectors become inoperable if someone turns the power off at the circuit breaker. The solution in this situation is to turn the circuit breaker back on.

Most smoke alarms containing batteries will signal a low-battery condition by emitting a chirp every few seconds. This chirp indicates that the battery needs to be replaced to keep the smoke alarm functional. Many hard-wired smoke alarms that are equipped with battery backups will chirp to indicate that the backup battery is low, even when the 110-volt power source is operational. This feature assures that the smoke alarm will always have an operational backup system.

Dirt Problems

A second problem with smoke alarms is the increased sensitivity that results when dust or an insect becomes lodged in the photoelectric or ionization chamber. Such an obstruction causes the alarm to activate when a small quantity of water vapor or smoke enters the chamber, leading to unnecessary alarms. The solution in this situation is to remove the cover of the chamber and gently vacuum out the chamber. Follow the manufacturer's instructions for this process.

Alarm Problems

The third common problem you may encounter with smoke alarms is a worn-out detector. It is recommended that smoke alarms be replaced every 10 years because the sensitivity of the alarm can change. Some worn-out alarms may become overly sensitive and emit false alarms. Others may develop decreased sensitivity and fail to emit an alarm in the event of a fire. The date of manufacture should be stamped on each smoke alarm. If a detector is more than 10 years old, it should be replaced with a new alarm.

Understanding the basic functions and troubleshooting of simple smoke alarms enables you to respond to citizens who call your fire department when they encounter problems with their smoke alarms. Most importantly, it enables you to keep smoke alarms operational. Remember—only *operational* smoke alarms can save lives.

Smoke detectors come in a variety of designs and styles geared toward different applications. The most common smoke detectors are ionization and photoelectric models, which operate in the same way that residential smoke alarms do. However, the smoke detectors used in commercial fire alarm systems are much more sophisticated and more expensive than residential smoke alarms. In commercial fire alarm systems, photoelectric detectors are typically used, as ionization detectors have a tendency to trigger false alarms more frequently.

Each detection device is rated to protect a certain floor area, so in large areas the detectors are often placed in a grid pattern. Newer smoke detectors also have a visual indicator, such as a steady or flashing light, that indicates when the device has power or has been activated.

A **beam detector** is a type of photoelectric smoke detector used to protect large spans such as churches, auditoriums, airport terminals, and indoor sports arenas. In these facilities, it would be difficult or costly to install large numbers of individual smoke detectors, but a single beam detector can be used for the entire length (**FIGURE 6-12**).

A typical beam detector has two components: a sending unit, which projects a narrow beam of light across the open area, and a receiving unit, which measures the intensity of the light when the beam strikes the receiver. When smoke interrupts the light beam, the receiver detects a drop in the light intensity and activates the fire alarm system. Most photoelectric beam detectors are set to respond to a certain **obscuration rate**, or percentage of light blocked. If the light is completely blocked, as when a solid object

FIGURE 6-12 Beam detectors are used in large open spaces.

Courtesy of Fire Fighting Enterprises, Ltd.

is moved across the beam, the trouble alarm will sound, but the fire alarm will not be activated.

Smoke detectors are usually powered by a low-voltage circuit and send a signal to the fire alarm control panel when they are activated. Both ionization and photoelectric smoke detectors are self-restoring. After the smoke condition clears, the alarm system can be reset at the control panel.

Heat Detectors. **Heat detectors** are also commonly used as automatic alarm initiation devices. Heat detectors can provide property protection, but cannot provide reliable life-safety protection because they do not react quickly enough to incipient fires. Because of this, the detectors have written on them "not a life safety device." They are generally used in situations where smoke alarms cannot be used, such as dusty environments and areas that experience extreme cold or heat. These detectors are often installed in unheated areas, such as attics and storage rooms, as well as in boiler rooms and manufacturing areas.

Heat detectors are generally very reliable and less prone to false alarms than are smoke detectors. You may come across heat detectors that were installed 30 or more years ago and are still in service; however, some older units have no visual trigger that tells which device was activated, so tracking down the cause of an alarm may be very difficult. Newer models have an indicator light that shows which device was activated.

Several types of heat detectors are available, each of which is designed for specific situations and applications. **Spot detectors** are individual units that can be spaced throughout an occupancy, so that each detector covers a specific floor area. The detectors may be in individual rooms or spaced at intervals along the ceiling in larger areas.

Heat detectors can be designed to operate at a fixed temperature or to react to a rapid increase in temperature. Either fixed-temperature or rate-of-rise devices can be configured as spot or line detectors.

Fixed-Temperature Heat Detectors. **Fixed-temperature heat detectors**, as the name implies, are designed to operate at a preset temperature. A typical temperature for a light-hazard occupancy, such as an office building, would typically be 135°F (57°C); however, they can be provided in other activation temperatures, depending on the locations where these will be installed. Fixed-temperature detectors typically include a metal alloy that will melt at the preset temperature. The melting alloy releases a lever-and-spring mechanism, to open or close a switch. Most fixed-temperature heat detectors must be replaced after they have been activated, even if the activation was accidental.

Rate-of-Rise Heat Detectors. **Rate-of-rise heat detectors** will activate if the temperature of the surrounding air rises more than a set amount in a given period of time. A typical rating might be "greater than 12°F (6.7°C) in 1 minute." If the temperature increase occurs more slowly than this rate, the rate-of-rise heat detector will not activate. By contrast, a temperature increase at a pace greater than this rate will activate the detector and set off the fire alarm. Rate-of-rise heat detectors should not be located in areas that normally experience rapid changes in temperature, such as near garage doors in heated parking areas, heating registers, or a commercial kitchen's cooking line.

Some rate-of-rise heat detectors have a **bimetallic strip** made of two metals that respond differently to heat: A rapid increase in temperature causes the strip to bend unevenly, which opens or closes a switch. Another type of rate-of-rise heat detector uses an air chamber and diaphragm mechanism: As air in the chamber heats up, the pressure increases. Gradual increases in pressure are released through a small hole, but a rapid increase in pressure will press upon the diaphragm and activate the alarm. Most rate-of-rise heat detectors are self-restoring, so they do not need to be replaced after an activation unless they were directly exposed to a fire.

Rate-of-rise heat detectors generally respond more rapidly to most fires than do fixed-temperature heat detectors. However, a slow-burning fire, such as a smoldering couch, may not activate a rate-of-rise heat detector until the fire is well established. Combination rate-of-rise and fixed-temperature heat detectors are available. These devices balance the faster response of the rate-of-rise detector with the reliability of the fixed-temperature heat detector.

Line Heat Detectors. **Line detectors** use wire or tubing strung along the ceiling of large open areas to detect an increase in heat. An increase in temperature anywhere along the line will activate the detector. Line detectors are found in churches, warehouses, and industrial or manufacturing applications.

One wire-type model has two wires inside, separated by an insulating material. When heat melts the insulation, the wires short out and activate the alarm. The damaged section of insulation must be replaced with a new piece after activation.

Another wire-type model measures changes in the electrical resistance of a single wire as it heats up. This device is self-restoring and does not need to be replaced after activation unless it is directly exposed to a fire.

The tube-type line heat detector contains a sealed metal tube filled with air or a nonflammable gas. When the tube is heated, the internal pressure increases and activates the alarm. Like the single-wire line heat detector, this device is self-restoring and does not need to be replaced after activation unless it is directly exposed to a fire.

Flame Detectors. **Flame detectors** are specialized devices that detect the electromagnetic light waves produced by a flame (**FIGURE 6-13**). These devices can recognize a very small fire extremely quickly—as soon as a match is struck—and prior to the flame stabilizing on the match.

FIGURE 6-13 Flame detectors are specialized devices that detect the electromagnetic light waves produced by a flame.

Typically flame detectors are found in places such as aircraft hangars or specialized industrial settings in which early detection and rapid reaction to a fire are critical. Flame detectors are also used in explosion suppression systems, where they detect and suppress an explosion as it is occurring.

Flame detectors are complicated and expensive. Another disadvantage of these models is that other infrared or ultraviolet sources, such as the sun or a welding operation, can set off an unwanted alarm. Flame detectors that combine infrared and ultraviolet sensors are sometimes used to lessen the chances of a false alarm.

Gas Detectors. **Gas detectors** are calibrated to detect the presence of a specific gas that is created by combustion or that is used in the facility. Depending on the system, a gas detector may be programmed to activate either the building's fire alarm system or a separate alarm. These specialized instruments need regular calibration if they are to operate properly. Gas detectors are usually found only in specific commercial or industrial applications.

Air Sampling Detectors. **Air sampling detectors** continuously capture air samples and measure the concentrations of specific gases or products of combustion. These devices draw in air samples through a sampling unit and analyze them using an ionization or photoelectric smoke detector. When air sampling detectors are installed in the return air ducts of large buildings, they are known as **duct detectors**. They will sound an alarm and shut down the air-handling system if they detect smoke.

More complex systems are sometimes installed in special hazard areas to draw air samples from rooms, enclosed spaces, or equipment cabinets (**FIGURE 6-14**). The samples pass through gas analyzers that can identify smoke particles, products of combustion, and concentrations of other gases associated with a dangerous condition. Air sampling detectors are most often used in areas that hold valuable contents or sensitive equipment and, therefore, where it is important to detect problems early.

Alarm Initiation by Fire Suppression Systems. Other fire protection systems in a building may activate the fire alarm system. Automatic sprinkler systems are usually connected to the fire alarm system through a water flow paddle, and will activate the alarm if a water flow occurs (**FIGURE 6-15**). Dry pipe sprinkler systems use pressure switches to activate the fire alarm system. Such

FIGURE 6-14 Air sampling detector air intake cone.

FIGURE 6-15 Automatic sprinkler systems use an electric flow switch to activate the building's fire alarm system.

a system not only alerts the building occupants and the fire department to a possible fire, but also ensures that someone is made aware that water is flowing, in case of an accidental discharge. Any other fire-extinguishing systems in a building, such as those found in kitchens or containing halogenated or carbon dioxide agents, should also be tied into the building's fire alarm.

In addition to the water flow alarm, valves that can shut down the sprinkler system must have tamper switches on them. This is to indicate that someone is shutting down a valve that will affect the sprinkler system. In sprinklered buildings, this is a prime reason why trouble alarms are investigated by trained personnel.

PRO TIPS

Duct detectors are best powered by the control panel. Some duct detectors are powered by the air handling unit as they are often integral to the HVAC unit. The problem with this is when a maintenance person comes to work on the air handling unit they will shut the unit down. Doing so sends the fire alarm into trouble, as power to the detector has been lost.

Alarm Notification Appliances

Audible and visual alarm notification appliances such as bells, horns, and electronic speakers produce an audible signal when the fire alarm is activated. Most newer systems also incorporate visual alerting devices. These audible and visual alarms alert occupants of a building to a fire.

Older systems used a variety of sounds as notification devices and a building could have many different alarms ringing throughout the day. This inconsistency often led to confusion over whether the sound was actually an alarm. More-recent model building and fire prevention codes have adopted a standardized audio pattern, called the **temporal-3 pattern**, that must be produced by any audio device used as a fire alarm. The temporal 3 is three short sounds from the audible device then a longer pause, and then it repeats itself until the alarm is silenced. Even single-station smoke alarms designed for residential occupancies now use this sound pattern. As a consequence, people can recognize a fire alarm immediately.

Some public buildings also play a recorded evacuation announcement in conjunction with the temporal-3 pattern. The recorded message is played through the fire alarm speakers and provides safe evacuation instructions. In facilities such as airport terminals, this announcement is recorded in multiple languages. This kind of system may also include a public address feature that fire department or building security personnel can use to provide specific instructions, relay information about the situation, or give notice when the alarm condition is terminated.

Many new fire alarm systems incorporate visual notification devices such as high-intensity strobe lights as well as audio devices (**FIGURE 6-16**). Visual devices alert hearing-impaired occupants to a fire alarm and are very useful in noisy environments where an audible alarm might not be heard. Additionally, there are now requirements within the Fire Alarm Code (NFPA 72) that require low-frequency sounders in sleeping areas with new systems.

FIGURE 6-16 This alarm notification device has both a loud horn and a high-intensity strobe light.

TABLE 6-2 Categories of Alarm Annunciation Systems

Category	Description
Noncoded alarm	No information is given on what device was activated or where it is located.
Zoned noncoded alarm	Alarm system control panel indicates the zone in the building that was the source of the alarm. It may also indicate the specific device that was activated.
Zoned coded alarm	The system indicates over the audible warning device which zone has been activated. This type of system is often used in hospitals, where it is not feasible to evacuate the entire facility.
Master-coded alarm	The system is zoned and coded. The audible warning devices are also used for other emergency-related functions.

Other Fire Alarm Functions

In addition to alerting occupants and summoning the fire department, fire alarm systems may control other building functions, such as air-handling systems, fire doors, and elevators. To control smoke movement through the building, the system may shut down or start up air handling systems. Fire doors that are normally held open by electromagnets may be released to compartmentalize the building and confine the fire to a specific area. Doors allowing reentry from exit stairways into occupied areas may be unlocked. Elevators will be summoned to a predetermined floor, usually the main lobby, so they can be used by fire crews.

Responding fire personnel must understand which building functions are being controlled by the fire alarm, for both safety and fire suppression reasons. This information should be gathered during preincident planning surveys and should be available in printed form or on a graphic display at the control panel location.

Fire Alarm Systems

Some fire alarm systems give very little, or no, information at the alarm control panel; others specify exactly which initiation device activated the fire alarm. The systems can be further subdivided based on whether they are zoned or coded systems.

In a **zoned system**, the alarm control panel will indicate where in the building the alarm was activated. Almost all alarm systems are now zoned to some extent. Only the most rudimentary alarm systems give no information at the alarm control panel about where the alarm was initiated. In a **coded system**, the zone is identified not only at the alarm control panel but also through throughout the building using the audio notification device.

TABLE 6-2 shows how, using these two variables, systems can be broken down into four categories: noncoded alarm, zoned noncoded alarm, zoned coded alarm, and master-coded alarm.

Noncoded Alarm System

In a **noncoded alarm system**, the control panel has no information indicating where in the building the fire alarm was activated. The alarm typically sounds a bell or horn. Fire department personnel must then search the entire building to find which initiation device was activated. This type of system is generally found only in older, small buildings.

Zoned Noncoded Alarm System

The **zoned noncoded alarm system** is the most common type of system, particularly in smaller buildings. With this model, the building is divided into multiple zones, often by floor or by wing. The alarm control panel indicates in which zone the activated device is located and, sometimes, which type of device was activated. Responding personnel can go directly to that part of the building to search for the problem and check the activated device.

Many zoned noncoded alarm systems have an individual indicator light for each zone. When a device in that area is activated, the indicator light goes on.

Computerized alarm systems, known as addressable system, may use addressable devices. This is a step up from the zoned systems. In these systems, each individual initiation device—whether it is a smoke detector, heat detector, or pull-station—has its own unique identifier. When the device is activated, the display on the control panel will indicate a specific point. When programmed properly, the display will provide information such as, "smoke detector, room 101." Responding personnel can then quickly locate the device or devices that have been activated.

Zoned Coded Alarm

In addition to having all the features of a zoned alarm system, a **zoned coded alarm system** indicates which zone has been activated over the announcement system. Hospitals often use this type of system, because it is not possible to evacuate all staff and patients for every fire alarm. The audible notification devices sound in a sequence that provides a code to indicate which zone was activated. A code list tells building personnel which zone is in an alarm condition.

More modern systems in these occupancies use speakers as their alarm notification devices. This approach means that a voice message indicating the nature and location of the alarm can accompany the audible alarm signal.

Master-Coded Alarm

In a **master-coded alarm system**, the audible notification devices for fire alarms also are used for other purposes. For example, a school may use the same bell to announce a change in classes, to signal a fire alarm, to summon the janitor, or to make other notifications. Most of these systems have been replaced by modern speaker systems that use the temporal-3 pattern fire alarm signal and have public address capabilities. This type of system is rarely installed in new buildings.

Fire Department Notification

The fire department should always be notified when a fire alarm system is activated. In some cases, a person must make a telephone call to the fire department or an emergency communications center. In other cases, the fire alarm system may be connected directly to the fire department or to a remote location where someone on duty receives a signal from the fire alarm system and calls the fire department.

There are two primary ways fire alarm systems transmit their signals to the fire department: phone lines and radio signal. Phone lines are the traditional method. When using a dedicated phone line—one that is only for monitoring the fire alarm—the monthly costs can be high. Recently there has been a move to switch from phone lines to radio to transmit the alarm signal. Radios have far less related trouble signals than when using phone lines. Additionally if a radio goes bad, it could be replaced in about an hour. If the phone line is dug up, it may be many days before it is repaired, leaving the building without automatic fire department notification.

As shown in **TABLE 6-3**, fire alarm systems can be classified in five categories, based on how the fire department is notified of an alarm.

Protected Premises Fire Alarm System

A **protected premises fire alarm system** does not notify the fire department. Instead, the alarm sounds only in the building to notify the occupants. Buildings with this type of system should have notices posted requesting occupants to call the fire department and report the alarm after they exit (**FIGURE 6-17**). It is common that after a period of a local alarm ringing, the occupants will call the fire department and ask if they are responding. This may be the first indication to the fire department that the alarm has activated. Most people assume that when a fire alarm activates, it is only minutes until the arrival of the fire department. For the safety of occupants, it is highly suggested that these alarms transmit their signals to a monitoring station.

Remote Supervising Station Systems

A **remote supervising station system** sends a signal directly to the fire department or to another monitoring location via a telephone line or a radio signal. This type of direct notification can only be installed in jurisdictions where the fire department is equipped to handle direct alarms. If the signal goes to a monitoring location, that site must be continually staffed by someone who will call the fire department.

TABLE 6-3 Fire Department Notification Systems

Type of System	Description
Protected Premises	The fire alarm system sounds an alarm only in the building where it was activated. No signal is sent out of the building. Someone must call the fire department to respond.
Remote Supervising Station	The fire alarm system sounds an alarm in the building and transmits a signal to a remote location. The signal may go directly to the fire department or to another location where someone is responsible for calling the fire department.
Auxiliary System	The fire alarm system sounds an alarm in the building and transmits a signal to the fire department via a public alarm box system.
Proprietary Supervising System	The fire alarm system sounds an alarm in the building and transmits a signal to a monitoring location owned and operated by the facility's owner. Depending upon the nature of the alarm and arrangements with the local fire department, facility personnel may respond and investigate, or the alarm may be immediately retransmitted to the fire department. These facilities are monitored 24 hours a day.
Central Station	The fire alarm system sounds an alarm in the building and transmits a signal to an off-premises alarm monitoring facility. The off-premises monitoring facility is then responsible for notifying the fire department to respond, and they must send a person to investigate the signals and provide silence or reset services.

Auxiliary Systems

Auxiliary systems can be used in jurisdictions with a public fire alarm box system. A building's fire alarm system is tied into a master alarm box located outside the building. When the alarm activates, it trips the master box, which transmits the alarm directly to the fire department communications center. This would only be seen in large older cities, as most cities have phased out public fire alarm boxes.

FIGURE 6-17 Buildings with a local alarm system should post notices requesting occupants to call the fire department and report the alarm after they exit.

Proprietary Supervising Systems

In a **proprietary supervising system**, the building's fire alarms are connected directly to a monitoring site that is owned and operated by the building's owner. Proprietary systems are often installed at facilities where multiple buildings belong to the same owner, such as universities or industrial complexes. Each building is connected to a monitoring site on the premises (usually the security center), which is staffed at all times (**FIGURE 6-18**). When an alarm sounds, the staff at the monitoring site reports the alarm to the fire department, usually by telephone or a direct line. The code does not state that the buildings need to be on the same property. Take, for example, a large box retailer who owns buildings throughout the country. Because they are all owned by the same owner, they fall into the proprietary category and can have all building monitored in one location, which may be on the opposite side of the country from your jurisdiction.

Central Stations

A **central station** is a third-party, off-site monitoring facility that monitors multiple alarm systems. Individual building owners contract and pay the central station to monitor their facilities (**FIGURE 6-19**). When an activated alarm at a covered building transmits a signal to the central station by telephone or radio, personnel at the central station notify the appropriate fire department of the fire alarm. The central station facility may

FIGURE 6-18 In a proprietary system, fire alarms from several buildings are connected to a single monitoring site owned and operated by the buildings' owner.

be located in the same city as the facility or in a different part of the country. One of the requirements of a central station is to provide a "runner." This is someone who would respond to the building to investigate and repair the alarm if necessary.

Usually, building alarms are connected to the central station through leased or standard telephone lines; however, the use of either cellular telephone frequencies or radio frequencies is becoming more common. Cellular or radio connections may be used to back up regular telephone lines in case they fail; in remote areas without telephone lines, they may be the primary transmission method.

FIGURE 6-19 A central station monitors alarm systems at many locations.

Wiring Concerns

Generally speaking, correct wiring begins at one device and goes to the next, and the next, until that last device is connected on that circuit. This type of wiring is done to ensure the integrity of the wiring, and the devices are part of that integrity. When a device is removed from the circuit, or wiring is cut the panel sees a problem and reports a trouble alarm. Until that problem is corrected, some of the devices will not work. Everything up to the missing detector will function as the wiring circuit is still intact. The devices past the missing detector will not operate as the wiring has been "opened" because of the missing detector.

People not familiar with wiring fire alarms will add devices by **T tapping** into the existing wiring. Instead of properly wiring from one device into the next, they take a couple of wires and join them to the existing circuit. While the device will function if activated, the problem is that it is not **supervised**. That means that if the device is removed from the wiring, the control panel will not know, and no trouble alarm will be transmitted.

Fire Alarm and Detection System Maintenance

Fire alarm systems must be kept in proper working order for the system to work as designed. That means the system must be maintained properly and when a device is broken it must be repaired immediately. In addition, there should be a monthly visual inspection looking for things such as visibility of pull stations, access to the pull stations, smoke detectors with the caps still in place, damage to any devices, and anything else that might prohibit smoke or heat detectors from operating such as painted heat detectors and bags on

FIGURE 6-20 There should be a monthly visual inspection of fire detection systems looking for things such as visibility of pull stations, access to the pull stations, smoke detectors with the caps still in place, or damage to any devices.

smoke detectors (**FIGURE 6-20**). Also, yearly system tests should be conducted to determine that the system is operating as required. NFPA 72 specifies the requirements of inspections and maintenance.

One thing that is not visible is the strength of the batteries. They should be replaced as recommended by the manufacturer. Load testing during an inspection can determine exactly when the batteries should be replaced.

Fire Alarm and Detection System Testing

As a first responder inspector, you should not conduct any tests on fire detection systems, but rather call for the owner to have the system inspected by an alarm company yearly. You will then study the inspection results to determine if action needs to be taken. You need to read this report closely. Often deficiencies are noted and recommended for correction. Unless you follow up with the owner, there is no way to know if repairs have actually been made to return the alarm to full operational status.

When a system is first installed, every part of the system should be tested.

The fire alarm and detection system should be designed to work on battery power, so for the test the 110 volt power should be shut off. This is the only way to tell of the audio visual devices have enough power to operate properly. Each device component of the system needs to be activated to see that it functions as designed. If auxiliary functions such as releasing magnetically held doors, or shutting down an HVAC unit are programmed, those must be checked as well to see that those functions occurred.

If there is an interruption in the wiring, such as a missing detector, a trouble alarm should occur. The best way to check this is to actually take some devices off of each circuit and see if a trouble alarm sounds.

When inspecting audio visual devices, it should be noted that for those people who might be susceptible to seizures the visual devices located within the space must flash together. They cannot each flash independently of each other.

During the inspection, you should ensure that proper signals go to the monitoring station. This is also a time to verify that the proper alarm sequence was received by the monitoring system. Each of the three types of alarms—trouble, supervisory, and fire—are increasingly more important. A properly functioning alarm system will have each type of alarm override the lower level alarm.

During the inspection, the circuit breaker needs to be identified and locked out at the breaker box to prevent an accidently shutting down of the fire alarm power. It is a good idea to also put the circuit number in the control panel for future use. The batteries in the control panel, and any power supplies in the field, must be marked with the year that they were manufactured.

After a fire inspector has accepted the system as installed correctly and functioning, unless the system is modified, not need to be this physically involved in the annual inspection process. As the first responder inspector, your job entails understanding the components, and determining the operational readiness of the system. It is critical to be able to read and interpret the inspection reports from the alarm contractor to ensure an operational state.

Documentation and Reporting of Issues

When conducting the inspection, any deficiencies need to be noted. There should be a copy of the results for you, the alarm company, and the owner. If you have to return, then only those items need to be inspected again. If the system meets your approval, then simply indicate on the paperwork that the system is approved. Often the alarm company will have an inspection report that they use. This paperwork should be kept in the occupancy file.

Annual inspection paperwork should be submitted to the owner by an alarm company indicating the status of the system. If deficiencies are noted, continued follow up must occur until the proper repairs a made. The most recent inspection paperwork needs to be kept on site at the control panel. Some alarm companies are installing tubes or boxes to house that paperwork.

WRAP-UP

CHAPTER SUMMARY

- Fire alarm and detection systems can protect the occupants and the structure by providing early warning to both the occupants and firefighters, which can minimize both loss of life and property damage during fires.
- A fire alarm and detection system recognizes when a fire is occurring and activates the fire alarm system, which then alerts the building occupants, activates building functions such as recalling elevators, shutting down HVAC units, unlocking doors, releasing door hold open devices, and in most cases, notifying the fire department.
- A fire detection or alarm system has three basic components: an alarm initiation device, an alarm notification device, and a control panel.
- The alarm notification appliance is generally an audible device, more recently accompanied by a visual indication, that alerts the building occupants once the system is activated.
- The control panel links the initiation device to the notification appliance and performs other essential functions as required.
- Two types of fire detection devices may be used in a smoke alarm to detect combustion:
  - Ionization detectors are triggered by the invisible products of combustion.
  - Photoelectric detectors are triggered by the visible products of combustion.
- Alarm initiation devices begin the fire alarm process either manually or automatically. Manual initiation devices require human action; automatic devices function without human intervention.
- Automatic initiation devices can use any of several types of detectors. Some detectors are activated by smoke; others react to heat, others to the light produced by actual flame, and still others to specific gases.
- Audible and visual alarm notification devices such as bells, horns, and electronic speakers produce an audible signal when the fire alarm is activated. Most new systems also incorporate visual alerting devices.
- In addition to alerting occupants and summoning the fire department, fire alarm systems may control other building functions, such as air-handling systems, fire doors, and elevators.
- Some fire alarm systems give very little, or no, information at the alarm control panel; others specify exactly which initiation device activated the fire alarm. The systems can be further subdivided based on whether they are zoned or coded systems.
- There are two ways fire alarm systems transmit their signals to the fire department—phone lines and radio signal.
- Correct wiring begins at one device and goes to the next, and the next, until that last device is connected on that circuit. This type of wiring is done to ensure the integrity of the wiring, and the devices are part of that integrity. When a device is removed from the circuit, or wiring is cut the panel sees a problem and reports a trouble alarm.
- Fire alarm systems must be kept in proper working order for the system to work as designed. That means the system must be maintained properly and when a device is broken it must be repaired immediately. In addition, there should be a monthly visual inspection.
- As a first responder inspector, you should not conduct any tests on fire detection systems, but rather call for the owner to have the system inspected by an alarm company yearly. You will then study the inspection results to determine if action needs to be taken.

KEY TERMS

Air sampling detector A system that captures a sample of air from a room or enclosed space and passes it through a smoke detection or gas analysis device.

Alarm initiation device An automatic or manually operated device in a fire alarm system that, when activated, causes the system to indicate an alarm condition.

Alarm matrix A chart showing what will happen with the fire alarm system when an initiating device is activated.

Alarm notification appliance An audible and/or visual device in a fire alarm system that makes occupants or other persons aware of an alarm condition.

Auxiliary system A fire alarm system that sounds an alarm in the building and transmits a signal to the fire department via a public alarm box system.

Beam detector A smoke detection device that projects a narrow beam of light across a large open area from a sending unit to a receiving unit. When the beam is interrupted by smoke, the receiver detects a reduction in light transmission and activates the fire alarm.

Bimetallic strip A device with components made from two distinct metals that respond differently to heat. When heated, the metals will bend or change shape.

Central station An off-premises facility that monitors alarm systems and is responsible for notifying the fire department of an alarm. These facilities may be geographically located some distance from the protected building(s).

Coded system A fire alarm system design that divides a building or facility into zones and has audible notification devices that can be used to identify the area where an alarm originated.

Double-action pull-station A manual fire alarm activation device that requires two steps to activate the alarm. The person must push in a flap, lift a cover, or break a piece of glass before activating the alarm.

Duct detector A smoke detector inside an air handling unit that will sound an alarm and shut down the air handling unit when smoke is detected.

Fire alarm control panel The component in a fire alarm system that controls the functions of the entire system.

Fixed-temperature heat detector A sensing device that responds when its operating element is heated to a predetermined temperature.

Flame detector A sensing device that detects the radiant energy emitted by a flame.

Gas detector A device that detects and/or measures the concentration of dangerous gases.

Heat detector A fire alarm device that detects abnormally high temperature, an abnormally high rate-of-rise in temperature, or both.

Ionization smoke detector A device containing a small amount of radioactive material that ionizes the air between two charged electrodes to sense the presence of smoke particles.

Line detector Wire or tubing that can be strung along the ceiling of large open areas to detect an increase in heat.

Manual pull-station A device with a switch that either opens or closes a circuit, activating the fire alarm.

Master-coded alarm An alarm system in which audible notification devices can be used for multiple purposes, not just for the fire alarm.

Noncoded alarm An alarm system that provides no information at the alarm control panel indicating where the activated alarm is located.

Obscuration rate A measure of the percentage of light transmission that is blocked between a sender and a receiver unit.

Photoelectric smoke detector A device to detect visible products of combustion using a light source and a photosensitive sensor.

Proprietary supervising system A fire alarm system that transmits a signal to a monitoring location owned and operated by the facility's owner.

Protected premises fire alarm system A fire alarm system that sounds an alarm only in the building where it was activated. No signal is sent out of the building.

Rate-of-rise heat detector A fire detection device that responds when the temperature rises at a rate that exceeds a predetermined value.

Remote annunciator A secondary fire alarm control panel in a different location than the main alarm panel; it is usually located near the front door of a building.

Remote supervising station system A fire alarm system that sounds an alarm in the building and transmits a signal to the fire department or an off-premises monitoring location.

Single-action pull-station A manual fire alarm activation device that takes a single step—such as moving a lever, toggle, or handle—to activate the alarm.

Single-station smoke alarm A single device usually found in homes that detects visible and invisible products of combustion and sounds an alarm.

Smoke detector A device that detects smoke and sends a signal to a fire alarm control panel.

Spot detector A single heat-detector device; these devices are often spaced throughout an area.

Supervised Electronically monitoring the alarm system wiring for an open circuit

KEY TERMS CONTINUED

T tapping Improper wiring of an initiating device so that it is not supervised.

Temporal-3 pattern A standard fire alarm audible signal for alerting occupants of a building.

Zoned coded alarm A fire alarm system that indicates which zone was activated both on the alarm control panel and through a coded audio signal.

Zoned noncoded alarm A fire alarm system that indicates the activated zone on the alarm control panel.

Zoned system A fire alarm system design that divides a building or facility into zones so that the area where an alarm originated can be identified.

You Are the First Responder Inspector

You are inspecting an eight-story mixed-use occupancy. The building contains commercial occupancies on the first floor and residential condominiums on the upper floors. The building is about 40 years old. It is equipped with multiple smoke detectors in each residential unit. The first floor and the hallways and lobbies of the upper floors are protected by sprinklers. The fire alarm system is divided into multiple zones. The alarm system is monitored by a central station, which is responsible for contacting the fire department in the event of an alarm.

1. How would you classify the fire alarm system in this building?
   - A. A noncoded alarm system
   - B. A zoned noncoded alarm system
   - C. A zoned coded alarm system
   - D. A master-coded alarm system
2. Which type of notification does this alarm system use?
   - A. Local alarm system
   - B. Remote station system
   - C. Auxiliary system
   - D. Central station
3. If this building were new, which test would you perform on this system?
   - A. The central station alarm test
   - B. The noncoded alarm activation test
   - C. The auxiliary system test
   - D. None
4. As part of the maintenance plan, fire detection systems should be tested:
   - A. weekly.
   - B. monthly.
   - C. yearly.
   - D. daily.

CHAPTER

7

Fire Suppression Systems

NFPA 1030 THAT INCLUDES CHAPTER 6: FIRST RESPONDER INSPECTOR (NFPA 1031)

First Responder Inspector

- 6.5.6

ADDITIONAL NFPA STANDARDS

- **NFPA 12**, *Standard on Carbon Dioxide Extinguishing Systems*
- **NFPA 12A**, *Standard on Halon 1301 Fire Extinguishing Systems*
- **NFPA 13**, *Standard for the Installation of Sprinkler Systems*
- **NFPA 13D**, *Standard for the Installation of Sprinkler Systems in One and Two Family Dwellings and Manufactured Homes*
- **NFPA 13R**, *Standard for the Installation of Sprinkler Systems in Low-Rise Residential Occupancies*
- **NFPA 14**, *Standard for the Installation of Standpipes and Hose Systems*
- **NFPA 15**, *Standard for Water Spray Fixed Systems for Fire Protection*
- **NFPA 16**, *Standard for the Installation of Foam-Water Sprinkler and Foam-Water Spray Systems*
- **NFPA 17**, *Standard for Dry Chemical Extinguishing Systems*
- **NFPA 17A**, *Standard for Wet Chemical Extinguishing Systems*
- **NFPA 25**, *Standard for the Inspection, Testing, and Maintenance of Water-Based Fire Protection Systems*
- **NFPA 750**, *Standard on Water Mist Fire Protection Systems*
- **NFPA 1963**, *Standard for Fire Hose Connections*
- **NFPA 2001**, *Standard on Clean Agent Fire Extinguishing Systems*

KNOWLEDGE OBJECTIVES

After studying this chapter, you will be able to:

1. Describe dry-barrel and wet-barrel hydrants.
2. Identify the six types of sprinkler systems.
3. Identify the types of sprinkler heads.
4. Identify the three types of standpipes.
5. Describe how to verify the operational readiness of fire suppression systems.

SKILLS OBJECTIVES

1. Determine the operational readiness of an existing fixed fire suppression system.
2. Review test documentation and field observations to determine the operational readiness of an existing fixed fire system.
3. Recognize the need for corrective action with an existing fixed fire suppression system.

You Are the First Responder Inspector

As a new first responder inspector, you are stationed in a mostly residential area and have only a few office buildings with automatic sprinkler systems. During your training, you gained a basic familiarity of the operation of fire suppression systems but you feel far from being an expert. Technical information on fire suppression systems is readily available and you also find that it is continually changing.

1. What can you do to gain real-life experience in these systems?
2. Would it be practical and possible for you to contact a local area large fire department, a fire service fire prevention association, a state fire training school, a specialized fire insurance company, or take courses through the NFPA to learn more about sprinkler systems?
3. Is there a technical group that would allow you to sit in on meetings or training sessions?

Introduction

Fire suppression systems include automatic sprinkler systems, standpipe systems, and specialized extinguishing systems such as dry chemical systems. Most newly constructed commercial buildings incorporate at least one of these systems, and increasing numbers of residential dwellings are being built with residential sprinkler systems as well. Some of the codes you will need to consult when working with the systems discussed in this chapter are:

- NFPA 12, *Standard on Carbon Dioxide Extinguishing Systems*
- NFPA 13, *Standard for the Installation of Sprinkler Systems*
- NFPA 13D, *Standard for the Installation of Sprinkler Systems in One and Two Family Dwellings and Manufactured Homes*
- NFPA 13R, *Standard for the Installation of Sprinkler Systems in Low and Mid-Rise Residential Occupancies*
- NFPA 14, *Standard for the Installation of Standpipes and Hose Systems*
- NFPA 15, *Standard for Water Spray Fixed Systems for Fire Protection*
- NFPA 16, *Standard for the Installation of Foam-Water Sprinkler and Foam-Water Spray Systems*
- NFPA 17, *Standard for Dry Chemical Extinguishing Systems*
- NFPA 17A, *Standard for Wet Chemical Extinguishing Systems*
- NFPA 25, *Standard for the Inspection, Testing, and Maintenance of Water-Based Fire Protection Systems*

Types of Fire Hydrants

The fire department accesses the public or private water supply via fire hydrants. Most fire hydrants consist of an upright steel casing (barrel) attached to the underground water distribution system. The two major types of fire hydrants in use today are the dry-barrel hydrant and the wet-barrel hydrant. Hydrants are equipped with one or more valves to control the flow of water through the hydrant. One or more outlets are provided to connect fire department hoses to the hydrant. These outlet nozzles are sized to fit the 2½″ (63.5 mm) or larger fire hoses used by the local fire department.

The fire department accesses the public or private water supply via fire hydrants. (**TABLE 7-1**)

PRO TIPS

NFPA 24, Standard for the Installation of Private Fire Service Mains and Their Appurtenances, recommends that fire hydrants be color-coded to indicate the water flow available from each hydrant at 20 psi (138 kPa). It is recommended that the top bonnet and the hydrant caps be painted according to the system in TABLE 7-1.

Wet-Barrel Hydrants

Wet-barrel hydrants are used in locations where temperatures do not drop below freezing. These hydrants always have water in the barrel and do not have to be drained after each use. Wet-barrel hydrants usually have separate valves that control the flow to each individual outlet (**FIGURE 7-1**).

TABLE 7-1 Fire Hydrant Colors

Class	Flow Available at 20 psi	Color
Class C	Less than 500 gpm (1893 L/min)	Red
Class B	500–999 gpm (1893–3784 L/min)	Orange
Class A	1000–1499 gpm (3785–5677 L/min)	Green
Class AA	1500 gpm and higher (5678 L/min)	Light blue

FIGURE 7-1 A wet-barrel hydrant has a separate valve for each outlet nozzle.

Courtesy of American AVK Company.

Dry-Barrel Hydrants

Dry-barrel hydrants are used in climates where temperatures can be expected to fall below freezing. The valve that controls the flow of water into the barrel of the hydrant is located at the base, below the frost line, to keep the hydrant from freezing (**FIGURE 7-2**). The length of

A.

B.

FIGURE 7-2 A dry-barrel hydrant **(A)** is controlled by an underground valve **(B)**.

A: Courtesy of American AVK Company; **B:** © Jones & Bartlett Learning

FIGURE 7-3 Most dry-barrel hydrants have only one large valve at the bottom of the barrel that controls the flow of water.

the barrel depends on the climate and the depth of the valve. Water enters the barrel of the hydrant only when it is needed. Turning the nut on the top of the hydrant rotates the operating stem, which opens the valve so that water flows up into the barrel of the hydrant.

Whenever this type of hydrant is not in use, the barrel must remain dry. If the barrel contains standing water, it will freeze in cold weather and render the hydrant inoperable. After each use, the water drains out through an opening at the bottom of the barrel. This drain is fully open when the hydrant valve is fully shut. When the hydrant valve is opened, the drain closes, thereby preventing water from being forced out of the drain when the hydrant is under pressure.

A partially opened valve means that the drain is also partially open, so pressurized water can flow out. This leakage can erode (undermine) the soil around the base of the hydrant and may damage the hydrant. For this reason, a hydrant should always be either fully opened or fully closed. Most dry-barrel hydrants contain only one large valve that controls the flow of water (**FIGURE 7-3**). Each outlet nozzle must be connected to a hose or an outlet valve, or have a hydrant cap firmly in place before the valve is turned on.

Fire Hydrant Locations

Fire hydrants are located according to local standards and nationally recommended practices. Fire hydrants may be placed a certain distance apart, perhaps every 500 ft (152 m) in residential areas and every 300 ft (91 m) in high-value commercial and industrial areas. Consult your local community codes and standards. In many communities, hydrants are located at every street intersection, with mid-block hydrants being installed if the distance between intersections exceeds a specified limit.

A builder may be required to install additional hydrants when a new building is constructed due to local codes. For example, there may be a local code that requires that no part of the building will be more than a specified distance from the closest hydrant or a requirement that the hydrant cannot be more the X feet away from the fire department connection to the sprinkler system.

Automatic Sprinkler Systems

The most common type of fire suppression system is the **automatic sprinkler system**. Automatic sprinklers are reliable and effective, with a history of more than 100 years of successfully controlling fires. Unfortunately, few members of the general public have an accurate understanding of how automatic sprinklers work. In movies and on television, when one sprinkler head is activated, the entire system begins to discharge water. This inaccurate portrayal of how automatic sprinkler systems operate has made people hesitant to install automatic sprinklers, fearing that they will cause unnecessary water damage.

The reality is quite different. In almost all types of automatic sprinkler systems, the sprinkler heads open as each one is heated to its operating temperature. Usually, only one or two sprinkler heads open before the fire is controlled. There are some exceptions to this rule based upon sprinkler system design for a specific hazard.

A sprinkler system is a network of pipes that run underground and overhead (above ground). The underground pipes are connected to the building's water supply and supply water to the sprinkler system automatically. A valve in the sprinkler system riser or at the junction of the piping to the exterior water supply and the interior sprinkler system piping controls the water supply to the sprinkler system. The piping brings the water to the sprinkler heads (**FIGURE 7-4**).

Water Supply

Every automatic sprinkler system must have a water supply of adequate volume, pressure, and reliability. The types of hazards being protected, occupancy classification, and fuel loading conditions determine the minimum water flow required for the system. A water supply must be able to deliver the required volume of water to the highest or most remote sprinkler in a structure while maintaining a minimum residual pressure in the sprinkler system. Sprinkler systems must have a primary water supply and may be required to have a secondary source.

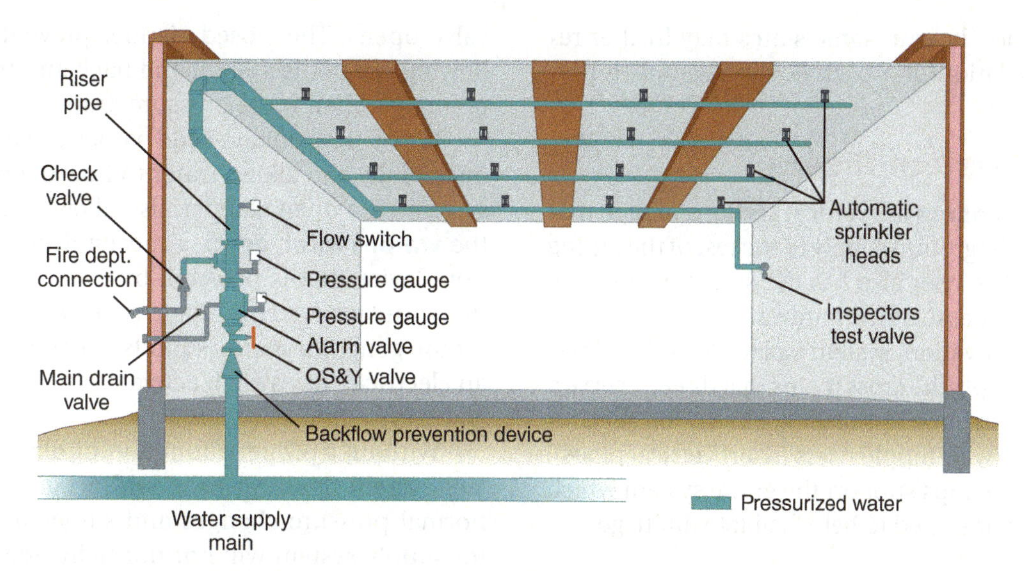

FIGURE 7-4 The basic components of an automatic sprinkler system include sprinkler heads, piping, control valves, and a water supply.

FIGURE 7-5 A fire department connection.

A **fire department connection (FDC)**, which is a fire hose connection through which the fire department can pump water into a sprinkler system or standpipe system, is also a necessary part of an automatic sprinkler system in order to ensure that adequate water pressure and volume supplied to the system are maintained (**FIGURE 7-5**). A fire department pumper connected to the public water supply by way of a fire hydrant can pump water into the sprinkler system vertical pipe section (riser) through the fire department connection.

Water Distribution Pipes

An automatic sprinkler system consists of an arrangement of pipes in different sizes. The system starts with an underground water supply water main (public or private). The underground supply water main contains a **check valve**, a valve that allows flow in one direction only, to prevent sprinkler system water from backflowing into the public water system and possibly contaminating the public drinking water supply. Water quality protection laws in some states may further require the installation of a back flow preventer to prevent backflowing.

Sprinkler System Risers

Sprinkler system **risers** are vertical sections of pipe that connect the underground supply to the rest of the piping in the system. The riser also has the system water flow control valve and associated hardware that is used for testing, alarm activation, system isolation, and maintenance. Risers supply the cross mains that directly serve a number of branch lines. Sprinkler heads are installed on the branch lines with nipple risers (short vertical pipes). Hangers, rings, clamps support the entire system which may be pitched or sloped to help facilitate drainage.

Valves

In order to control the water supply, a sprinkler system includes several different valves, such as the main water supply control valve, the alarm valve, and other smaller valves used for testing and service. These smaller valves include check valves, drain valves, globe valves, and alarm valves. Many large systems have zone valves, which enable the water supply to different areas to be shut down without turning off the entire system. All of the valves play a critical role in the design and function of the system.

PRO TIPS

Sprinkler contractors may also use a grid design or looped system design. A grid system design allows water to access the sprinkler heads from multiple directions. The looped system also allows water to come from multiple directions; however, the branch lines are not interconnected like they are in a grid system.

The type of main sprinkler system valve used depends on the type of sprinkler system installed. Options include an **alarm valve**, a **dry-pipe valve**, or a **deluge valve**. These valves are usually installed on the main riser, above the water supply control valve.

The primary functions of an alarm valve are to signal an alarm when a sprinkler head is activated and to prevent nuisance alarms caused by pressure variations and surges in the water supply to the system. The alarm valve has a clapper mechanism that remains in the closed position until a sprinkler head or inspector's test valve opens. The closed clapper prevents water from flowing out of the system and back into the public water mains when water pressure drops.

When a sprinkler head is activated, the clapper opens fully and allows water to flow freely through the system. The open clapper also allows water to flow to the **water-motor gong**, a mechanical alarm notification device that is powered by water moving through the sprinkler system. When water reaches the water-motor gong, an alarm sounds. In some installations, an electrical flow switch activates internal and external alarm systems.

Without a properly functioning alarm valve, sprinkler system flow alarms would occur frequently. The normal pressure changes and surges in a public water supply system will not normally open the clapper valve. The clapper valve prevents water from flowing to the water-motor gong or tripping the electrical water flow switches as a result of the pressure surges and changes in the public water supply system. In addition, some sprinkler systems may require the installation of a retard chamber due to severe water pressure changes in order to prevent the accidental tripping of the alarm valve.

Many of the newer wet pipe sprinkler systems do not use an alarm valve. Instead, they use a paddle-type water flow switch device that detects the movement of water and sounds the alarm. These paddle type devices use a dial device that can be adjusted from seconds to minutes to prevent a momentary surge of water pressure from activating the alarm.

In dry-pipe and deluge systems, the main valve functions both as an alarm valve and as a dam, holding back the water until the sprinkler system is activated. When the system is activated, the valve opens fully so that water can enter the **sprinkler piping**, which is the network of piping in a sprinkler system that delivers water to the sprinkler heads. Dry-pipe and deluge systems are described later in this chapter.

Main Water Supply Control Valves. Every sprinkler system must have at least one main control valve that allows water to enter the system. This water supply control valve must be of the "indicating" type, meaning that the position of the valve itself indicates whether it is open or closed. Examples are the **outside stem and yoke (OS&Y) valve**, the **post indicator valve (PIV)**, the **wall post indicator valve (WPIV)**, and the **butterfly valve**.

The OS&Y valve has a stem that moves in and out as the valve is opened or closed (**FIGURE 7-6**). If the stem is out, the valve is open. OS&Y valves are often found in

FIGURE 7-6 An outside stem and yoke (OS&Y) valve is often used to control the flow of water into a sprinkler system.

FIGURE 7-7 A post indicator valve (PIV) is used to open or close an underground valve.

a mechanical room in the building, where water to supply the sprinkler system enters the building. In warmer climates, they may be found outside.

The PIV has an indicator that reads either open or shut depending on its position (**FIGURE 7-7**). A PIV is usually located in an open area outside the building and controls an underground valve. Opening or closing a PIV requires a wrench, which is usually attached to the side of the valve.

A WPIV is similar to a PIV but is designed to be mounted on the outside wall of a building (**FIGURE 7-8**). The butterfly valve is a type of indicating valve, and when the butterfly valve is turned, there is a flat piece of

FIGURE 7-8 A WPIV controls the flow of water from an underground pipe into a sprinkler system.

metal inside the pipe that is rotated. When open, it is in line with the water flow and when closed, it is perpendicular to the flow. Determining if the butterfly valve is open or closed is easy, as there is an external indicator. When open, the indicator is in line with the piping and when closed, it is at a right angle to the piping.

It is critically important that the sprinkler system is always charged with water and ready to operate if needed. To ensure that the water supply to the sprinkler system is not deliberately or accidentally shut off, all control valves should be chained and locked in the open position, or electronically monitored with a **tamper switch** (**FIGURE 7-9**). Tamper switches electronically monitor the position of the valve. If someone opens or closes the valve, the tamper switch sends a supervisory signal to the fire alarm control panel, indicating a change in the valve position. This signal only sounds at the fire alarm control panel, and at the monitoring station if the alarms are sent off-site. This alarm does not sound all of the alarms in the building and should not initiate an emergency services response. If the change has not been authorized, the cause of the signal can be investigated and the problem can be corrected.

Sprinkler System Zoning

In large facilities, a sprinkler system may be divided into zones, where a specific valve or sprinkler riser controls the flow of water to a particular zone in an area of the building. The area of a building that a single sprinkler system riser can cover is regulated by

FIGURE 7-9 A tamper switch activates an alarm if someone attempts to close a valve that should remain open.

FIGURE 7-10 In tall buildings, a fire pump may be needed to maintain appropriate pressure in the sprinkler system.

NFPA 13, for example the maximum area for a single zone is 52,000 square feet (4831 m^2) for light and ordinary hazard occupancies. This is one reason that in a building's sprinkler room it would not be uncommon to see multiple risers next to each other, each protecting a particular area or zone of the building. This design in accordance with NFPA 13 also makes maintenance easy and can prove extremely valuable during firefighting operations.

After the fire is extinguished, sprinkler system water flow to the fire affected area can be shut off so that the activated sprinkler heads can be replaced. Fire protection in the rest of the building, however, is unaffected by the shutdown of the sprinklers in the fire damaged area.

Fire pumps are needed when the water comes from a static source. They may also be deployed to boost the pressure in some sprinkler systems, particularly for tall buildings (**FIGURE 7-10**). Because most municipal water supply systems do not provide enough pressure to control a fire on the upper floors of a high-rise building, fire pumps will turn on automatically when the sprinkler system activates or when the pressure drops to a preset level. In high-rise buildings, a series of fire pumps on upper floors may be needed to provide adequate pressure.

A large industrial complex could have more than one water source, such as a municipal system and a backup storage tank (**FIGURE 7-11**). Multiple fire pumps can provide water to the sprinkler and standpipe systems in different areas through underground pipes. Private hydrants may also be connected to the same underground system.

FIGURE 7-11 An elevated storage tank ensures that there will be sufficient water and adequate pressure to fight a fire.

PRO TIPS

Remember, fire pumps make pressure, not water. Some people think that when the water supply is insufficient for a building, they will just put in a fire pump. A pump will only provide more pressure to the already insufficient water supply.

Water Flow Alarms

All sprinkler systems should be equipped with a method for sounding an alarm whenever water begins flowing in the pipes. This type of warning is important both in case of an actual fire and in case of an accidental activation. Without these alarms, the occupants or the fire department might not be aware of the sprinkler activation. If a building is unoccupied, the sprinkler system could continue to discharge water long after a fire is extinguished, leading to extensive water damage.

Most systems incorporate a mechanical flow alarm called a water-motor gong (**FIGURE 7-12**). When the sprinkler system is activated and the main alarm valve opens, some water is fed through a pipe to a water-powered gong located on the outside of the building. This gong alerts people outside the building that there is water flowing. This type of alarm will function even if there is no electricity.

Accidental soundings of water-motor gongs are rare, but sometimes, a water surge will cause a momentary alarm. If a water-motor gong is sounding, water is probably flowing from the sprinkler system somewhere in the building. Fire companies that arrive and hear the distinctive sound of a water-motor gong know that there is a fire or that something else is causing the sprinkler system to flow water.

Most modern sprinkler systems also are connected to the building's fire alarm system by either an electric **flow switch** or a pressure switch. This connection will trigger the alarm to alert the building's occupants; a monitored system will notify the fire department or third-party monitoring company as well. Like water-motor gongs, flow and pressure switches can be accidentally triggered by water pressure surges in the system. To reduce the risk of accidental activations, these devices usually have a time delay before they will sound an alarm.

FIGURE 7-12 A water-motor gong sounds when water is flowing in a sprinkler system.

Types of Sprinkler Systems

There are four major classifications of automatic sprinkler systems. Each type of system includes a piping system that contains water from a source of supply to the sprinklers in the area under protection. The four major classifications are:

- Wet-pipe systems
- Dry-pipe systems
- Preaction systems
- Deluge systems

Although many buildings may use the same type of system to protect the entire facility, it is not uncommon to see two or three systems combined in one building. Some facilities use a wet sprinkler system to protect most of the structure, but implement a dry sprinkler or preaction system in a specific area. For example, the office area of a building will have a wet system installed, but the attic, which is unheated, will have a dry system, or even an anti-freeze system. Dry systems will also be common in large areas where the temperatures are below freezing, such as a cold storage warehouse.

Wet-Pipe Sprinkler Systems

A **wet-pipe sprinkler system** is the most common and least expensive type of automatic sprinkler system. As its name implies, the piping in a wet system is always filled with water. When a sprinkler head activates, water is immediately discharged onto the fire. The major drawback to a wet sprinkler system is that it cannot be used in areas where temperatures drop below freezing. Water will also begin to flow in such a system if a sprinkler head is accidentally opened or a leak occurs in the piping.

Larger unheated areas, such as a loading dock, can be protected with an antifreeze loop. An antifreeze loop is a small section of the wet sprinkler system that is filled with glycol or glycerin instead of water. A check valve separates the antifreeze loop from the rest of the sprinkler system. When a sprinkler head in the unheated area is activated, an antifreeze mixture sprays out first, followed by 100% water.

Dry-Pipe Sprinkler Systems

A **dry-pipe sprinkler system** operates much like a wet sprinkler system, except that the pipes are filled with pressurized air or nitrogen instead of water. A dry-pipe valve keeps water from entering the pipes until the air or nitrogen pressure drops below a threshold pressure. Dry systems are used in facilities, or areas, that may experience below-freezing temperatures, such as unheated warehouses, garages, or an unheated attic space.

The air or nitrogen pressure is set just enough to hold a clapper, which acts like a check valve, inside the dry-pipe valve in the closed position (**FIGURE 7-13**). When a sprinkler head opens, the air escapes. As the air pressure drops, the water pressure on the bottom side of the clapper forces the clapper open and water begins to flow into the pipes. When the water reaches the open sprinkler head, it is discharged onto the fire.

Dry sprinkler systems do not eliminate the risk of water damage from accidental activation. If a sprinkler head breaks, the air pressure will drop and water will flow, just as in a wet sprinkler system.

Dry sprinkler systems should have a high/low pressure alarm to alert building personnel if the system pressure changes. The activation of this alarm could mean one of two things: The compressor is not working or there is an air leak in the system. If the air pressure in the system is too low, the clapper will open and the system will fill with water. At that point, the system would essentially function as a wet sprinkler system, which could cause it to freeze in low temperatures. In this scenario, the system would have to be drained and reset to prevent the pipes from freezing.

Dry sprinkler systems must be drained after every activation so the dry-pipe valve can be reset. The clapper also must be reset, and the air pressure must be restored before the water is turned back on.

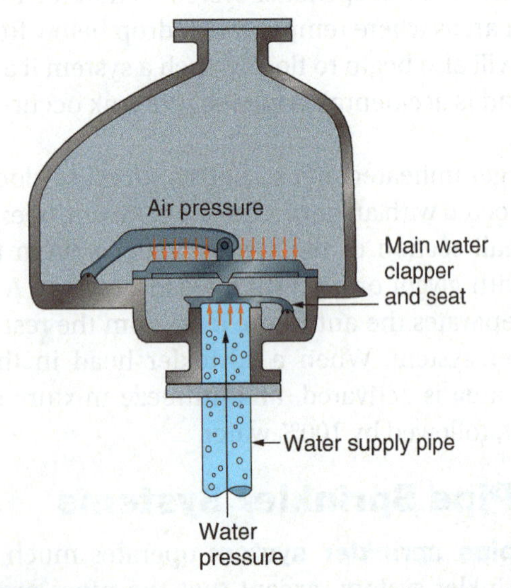

FIGURE 7-13 Water pressure on one side of the dry-pipe valve is balanced by air pressure on the other side of the valve.

Accelerators and Exhausters

One problem encountered in dry-pipe sprinkler systems is the delay between the activation of a sprinkler head and the actual flow of water out of the head. The pressurized air that fills the system must escape through the open head before the water can flow. For personal safety and property protection reasons, any delay longer than 90 seconds is unacceptable. Large systems, however, can take several minutes to empty out the air and refill the pipes with water. To compensate for this problem, two additional devices are used: accelerators and exhausters.

An **accelerator** is installed at the dry-pipe valve (**FIGURE 7-14**). The rapid drop in air pressure caused by an open sprinkler head triggers the accelerator, which allows air pressure to flow to the supply side of the clapper valve. The air is used to assist in opening the clapper. This quickly eliminates the pressure differential, opening the dry-pipe valve and allowing the water pressure to force the remaining air out of the piping.

An **exhauster** is installed on the system side of the dry-pipe valve, often at a remote location in the building. Like an accelerator, the exhauster monitors the air pressure in the piping. If it detects a drop in pressure, it opens a large-diameter portal, allowing the air in the pipes to escape to the atmosphere.

FIGURE 7-14 An accelerator is installed at the dry-pipe valve.

The exhauster closes when it detects water, diverting the flow to the open sprinkler heads. Large systems may have multiple exhausters located in different sections of the piping.

Preaction Sprinkler Systems

A **preaction sprinkler system** is also known as an interlock system, and is similar to a dry sprinkler system with one key difference: In a preaction sprinkler system, a secondary device—such as a smoke detector or a manual-pull alarm—must be activated before water is released into the sprinkler piping. This type of a system is known a single interlock system. A double interlock system would require the activation of a detection device *and* the opening of a sprinkler head prior to filling the sprinkler pipes. Once the system is filled with water, it functions as a wet sprinkler system.

A preaction system uses a deluge valve instead of a dry-pipe valve. The deluge valve will not open until it receives a signal that an initiating device, such as a smoke detector or pull station, has been activated. Because a detection system usually will activate more quickly than a sprinkler system does, water in a preaction system will generally reach the sprinklers before a head is activated. The fused head will then cause the water to flow from the head.

The primary advantage of a preaction sprinkler system is its ability to prevent accidental water discharges. If a sprinkler head is accidentally broken or the pipe is damaged, the deluge valve will prevent water from entering the system. These systems are very expensive and complicated, and usually reserved for areas where accidental water from a sprinkler system would be devastating to the room's contents, such as libraries and museums.

Combined Dry-Pipe and Preaction Systems

A combined dry-pipe and precaution system includes the essential features of both types of systems. The piping contains air or nitrogen under pressure. A supplementary heat-detecting device opens the water control valve and an air exhauster at the end of the unheated water feed main. The system then fills with water and operates as a wet-pipe system. If the supplementary heat-detecting should fail, then the system will operate as a conventional dry-pipe system.

Deluge Sprinkler Systems

A **deluge sprinkler system** is a type of dry sprinkler system in which water flows from all the sprinkler heads as soon as the system is activated **FIGURE 15**. A deluge system does not have closed heads that open individually at the activation temperature; instead, all of the heads in a deluge system are always open.

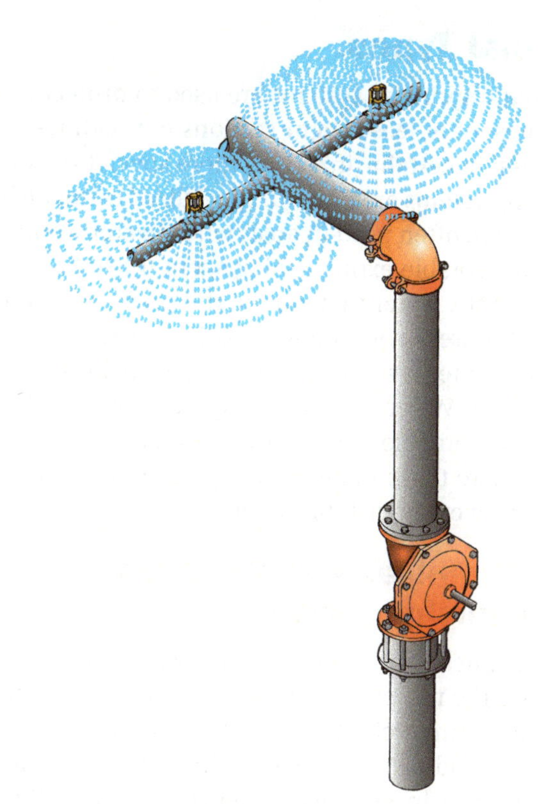

FIGURE 7-15 Water flows from all of the heads in a deluge system as soon as the system is activated.

Deluge systems can be activated in three ways:

1. A detection system can release the deluge valve when a detector is activated.
2. The deluge system can be connected to a separate pilot system of air-filled pipes with closed sprinkler heads. A water-filled system is known as a hydraulic, or wet, pilot system. When a head on the pilot line is activated, the air pressure drops, opening the deluge valve.
3. Most deluge valves can be released manually.

Deluge systems are used in special applications such as aircraft hangars or industrial processes, where rapid fire suppression is critical. In some cases, foam concentrate is added to the water, so that the system will discharge a foam blanket over the hazard. Deluge systems are also used for special hazard applications, such as liquid propane gas loading stations. In these situations, a heavy deluge of water is needed to prevent a large, rapidly-developing fire.

Special Types

Automatic sprinkler systems are used to protect whole buildings, or at least major sections of buildings. Nevertheless, in certain situations, more specialized extinguishing systems are needed. Specialized extinguishing systems are often used in areas where water would not be an acceptable extinguishing agent (**FIGURE 7-16**). For example, water may not be the preferred agent of choice for areas containing sensitive electronic equipment or contents such as computers, valuable books, or documents. Water is also incompatible with materials such as flammable liquids or water-reactive chemicals. Areas where these materials are present may require a specialized extinguishing system.

Dry Chemical and Wet Chemical Extinguishing Systems

Dry chemical and wet chemical extinguishing systems are the most common specialized agent systems. Used in commercial kitchens, they protect the cooking areas and exhaust systems. In addition, some gas stations have dry chemical systems that protect the dispensing areas. These systems are also installed inside buildings to protect areas where flammable liquids are stored or used. Both dry chemical and wet chemical extinguishing systems are similar in basic design and arrangement.

FIGURE 7-16 Special extinguishing systems are used in areas where water would not be effective or desirable.

PRO TIPS

Dry chemical extinguishing systems are not used as the agent of choice in commercial kitchens. When restaurants began to cook with vegetable oil and high-efficiency appliances, tests found that the standard dry chemical product would not extinguish the fire. While it would knock the fire down temporarily, the fire would quickly flash back to life. As a result, the NFPA created an additional fire classification, Class K, to address this specific hazard. Class K fires require the use of specialized extinguishing agents to extinguish a fire with cooking oils that have an extremely high combustion temperature. Chapter 8: Portable Fire Extinguishers discusses this topic in more detail.

Dry chemical extinguishing systems use the same types of finely powdered agents as dry chemical fire extinguishers (**FIGURE 7-17**). The agent is kept in self-pressurized tanks or in tanks with an external cartridge of carbon dioxide or nitrogen that provides pressure when the system is activated.

Five compounds that are used as the primary dry chemical extinguishing agents are:

- Sodium Bicarbonate—rated for class B and C fires only
- Potassium bicarbonate—rated for class B and C fires only

FIGURE 7-17 Dry chemical extinguishing systems are installed at many self-service gasoline filling stations.

- Urea-based potassium bicarbonate—rated for class B and C fires only
- Potassium Chloride—rated for class B and C fires only
- Ammonium phosphate—rated for Class A, B, and C fires

Wet chemical extinguishing systems discharge a proprietary liquid extinguishing agent. It is important to note that wet chemical extinguishing agents are not compatible with normal all-purpose dry chemical extinguishing agents. Only wet agents such as Class K, or B:C-rated dry chemical extinguishing agents should be used where these systems are installed.

All dry chemical extinguishing systems must meet the requirements of NFPA 17, *Standard for Dry Chemical Extinguishing Systems.* All wet chemical extinguishing systems must meet the requirements of NFPA 17A, *Standard for Wet Chemical Extinguishing Systems.* With both dry chemical and wet extinguishing agent systems, fusible-link or other automatic initiation devices are placed above the target hazard to activate the system (**FIGURE 7-18**). A manual discharge station is also provided so that workers can activate the system if they discover a fire (**FIGURE 7-19**). When the system is activated, the extinguishing agent flows out of all the nozzles. Nozzles are located over the target areas to discharge the agent directly onto a fire.

Many kitchen systems discharge agent into the ductwork above the exhaust hood as well as onto the cooking surface. This approach helps prevent a fire from igniting any grease buildup inside the ductwork and spreading throughout the system. Although the ductwork should be cleaned regularly, it is not unusual for a kitchen fire to extend into the exhaust system.

Dry and wet-chemical extinguishing systems should be tied into the building's fire alarm system.

FIGURE 7-18 Fusible links can be used to activate a specialized extinguishing system.

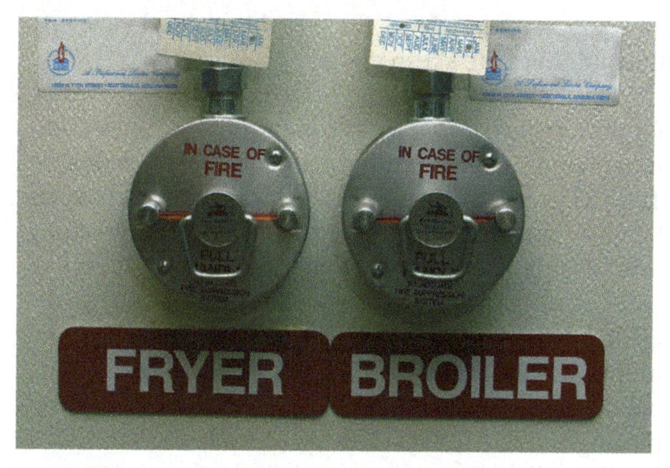

FIGURE 7-19 Most special extinguishing systems can also be manually activated.

Kitchen extinguishing systems should also shut down gas or electricity to the cooking appliances and exhaust fans.

Clean Agent Extinguishing Systems (Halogenated Agents)

Clean agent extinguishing systems are often installed in areas where computers or sensitive electronic equipment are used or where valuable documents are stored. The agents used in these systems are nonconductive and leave no residue. Halogenated agents or carbon dioxide are generally used for this purpose because they will extinguish a fire without causing significant damage to the contents.

Clean agent systems operate by discharging a gaseous agent into the atmosphere at a concentration that will extinguish a fire. Smoke detectors or heat detectors installed in these areas activate the system, although a manual discharge station is also provided with most installations. Discharge is usually delayed 30 to 60 seconds after the detector is activated to allow workers to evacuate the area.

During this delay (the pre-alarm period), an abort switch can be used to stop the discharge. In some systems, the abort button must be pressed until the detection system is reset; releasing the abort button too soon causes the system to discharge. If there is a fire, the clean agent system should be completely discharged before firefighters arrive.

Clean agent extinguishing systems should be tied to the building's fire alarm system and indicated as a zone on the control panel. This notification scheme alerts firefighters that they are responding to a situation where a clean agent has discharged. If the system has a preprogrammed delay, the pre-alarm should activate the building's fire alarm system.

Until the 1990s, Halon 1301 was the agent of choice for protecting areas such as computer rooms, telecommunications rooms, and other sensitive areas. **Halon 1301** is a nontoxic, odorless, colorless gas that leaves behind no residue. It is very effective at extinguishing fires because it interrupts the chemical reaction of combustion. Unfortunately, Halon 1301 damages the environment—a fact that led to a manufacturing and importation ban. Alternative agents have been developed for use in new types of systems that are also replacing Halon 1301 systems. A stockpile of Halon 1301 is still available to recharge existing systems, because the new clean agents are not compatible with the Halon system equipment.

Carbon Dioxide Extinguishing Systems

Carbon dioxide extinguishing systems are similar in design to clean agent systems. The primary difference is that carbon dioxide extinguishes a fire by displacing the oxygen in the room and smothering the fire. Large quantities of carbon dioxide are required for this purpose, because the area must be totally flooded to extinguish a fire (**FIGURE 7-20**).

Carbon dioxide systems may be designed to protect either a single room or a series of rooms. They usually have the same series of pre-alarms and abort buttons as are found in Halon 1301 systems. Because the carbon dioxide discharge creates an oxygen-deficient atmosphere in the room, the activation of this system is immediately dangerous to life. Any occupant who is still in the room when the agent is discharged is likely to be rendered unconscious and asphyxiated. Carbon dioxide extinguishing systems should be connected to the building's fire alarm system. Responding firefighters should see that a carbon dioxide system discharge has been activated. Using this knowledge, they can deal with the situation safely.

FIGURE 7-20 Carbon dioxide extinguishes a fire by displacing the oxygen in the room and smothering the fire.

PRO TIPS

NFPA 12A, *Standard on Halon 1301 Fire Extinguishing Systems*, and NFPA 2001, *Standard on Clean Agent Fire Extinguishing Systems*, provide the design and installation requirements for clean agent fire extinguishing systems.

Automatic Sprinkler Heads

Automatic sprinkler heads, which are commonly referred to as sprinkler heads, are the working ends of a sprinkler system. In most systems, the heads serve two functions: activate the sprinkler system and apply water to the fire. Sprinkler heads are composed of a body, or frame, which includes the orifice (opening); a release mechanism, which holds a cap in place over the orifice; and a deflector, which directs the water in a spray pattern (**FIGURE 7-21**). Standard sprinkler heads have a ½″ (12.7 mm) orifice, but several other sizes are available for special applications.

FIGURE 7-21 Automatic sprinkler heads have a body with an opening, a release mechanism, and a water deflector.

PRO TIPS

NFPA Standard 12, *Standard on Carbon Dioxide Extinguishing Systems*, provides the design and installation requirements for carbon dioxide extinguishing systems.

Types of Sprinkler Heads

The upright, the pendent, and the sidewall sprinkler head are basic sprinkler heads. Sprinkler heads with different mounting positions and temperature ratings are not interchangeable, because each mounting position has deflectors specifically designed to produce an effective water pattern down or out toward the fire. Each automatic sprinkler head is designed to be mounted in one of three positions. Additional sprinkler heads are variations of three basic sprinklers modified to address specific needs.

The **upright sprinkler head** are designed to be installed so that the water spray is directed upwards against the deflector (**FIGURE 7-22**). Upright sprinkler heads are designed to be mounted on top of the supply piping, as their name suggests. Upright heads are usually marked SSU, for "standard spray upright."

Pendent sprinkler heads are designed to be installed so that the water stream is directed downward against the deflector (**FIGURE 7-23**). Pendant sprinkler heads are designed to be mounted on the underside of the sprinkler piping, hanging down toward the room. These heads are commonly marked SSP, which stands for "standard spray pendant."

Sidewall sprinklers have special deflectors which are designed to discharge most of the water away from the nearby wall in a pattern resembling one quarter of a sphere, with a small portion of the discharge directed at the wall behind the sprinkler (**FIGURE 7-24**). Sidewall sprinkler heads are designed for horizontal mounting, projecting out from a wall. Extended coverage sidewall sprinklers offer special extended, directional, and discharge water patterns.

FIGURE 7-22 Upright sprinkler head.

FIGURE 7-23 Pendent sprinkler head.

FIGURE 7-24 Sidewall sprinkler head.

Dry sprinkler heads provide protection in an area where 40 degrees Fahrenheit cannot be maintained. They are constructed to provide isolation between the head and the water supply by use of a cylinder that extends from the head to the threads. The threads reside in a heated area or are installed on a dry sprinkler system that is designed with pendent heads.

Open sprinklers are sprinklers from which the actuating elements (fusible links), which trigger sprinkler activation, have been removed (**FIGURE 7-25**). Open sprinkler heads are used with deluge systems. **Corrosion-resistant sprinklers** have special coating or platings such as wax or lead and are used in atmospheres which could corrode an uncoated sprinkler. **Nozzle sprinklers** are used in sprinkler applications requiring special discharge patterns, directional spray, fine spray, or other unusual discharge characteristics (**FIGURE 7-26**). **Ornamental sprinklers** are sprinklers which have been painted or plated by the manufacturer. **Flush sprinklers** are sprinklers in which all or part of the body, including the shank thread, is mounted above the lower plane of the ceiling (**FIGURE 7-27**). **Recessed sprinklers** are sprinklers in which all or part of the body, other than the shank thread, is mounted within a recessed housing (**FIGURE 7-28**). **Intermediate level sprinklers** are sprinklers equipped with integral shields to protect their operating elements from the discharge of sprinklers installed at higher elevations (**FIGURE 7-29**).

FIGURE 7-25 Open sprinkler head.

FIGURE 7-26 Nozzle sprinkler head.

FIGURE 7-27 Flush sprinkler head.

FIGURE 7-28 Recessed sprinkler head.

FIGURE 7-29 Intermediate level sprinkler head.

FIGURE 7-31 The ESFR sprinkler head is larger than a standard sprinkler head and discharges about four to five times the amount of water.

FIGURE 7-30 Special sprinkler heads are used in residential systems. They open more quickly than commercial heads and discharge less water.

Residential sprinklers are sprinklers which have been specifically designed and listed for use in residential occupancies (**FIGURE 7-30**). Sprinkler heads that are intended for residential occupancies are manufactured with a release mechanism that provides a faster response. Residential heads usually have smaller orifices and release a limited flow of water, because they are used in small rooms with a limited fire load. The spray pattern of the water is also different than a typical head in that it must hit higher on the wall. Because of this, residential heads and standard heads are not interchangeable.

An **early-suppression fast-response (ESFR) sprinkler head** has improved heat collectors that speed up the response and ensure rapid release of water onto the fire (**FIGURE 7-31**). They are used in large warehouses and distribution facilities where early fire suppression is important. These heads often have large orifices to discharge large volumes of water.

A listed **large-drop sprinkler head** generates large drops of water of such size and velocity as to enable effective penetration of a high-velocity fire plume. The large-drop sprinkler head has a proven ability to meet prescribed penetration, cooling, and distribution criteria prescribed in the large drop sprinkler examination requirements.

Old style sprinklers direct only 40 to 60 percent of the total water initially in a downward direction and are designed to be installed with the deflector either in the upright or pendent position. The primary difference between the old style sprinkler, which is designed for installations in either the upright or pendent position, and the current upright and pendent sprinklers is the design of the deflector.

Release Temperature

Sprinkler heads are also rated according to their release temperature. The release mechanisms hold the cap in place until the release temperature is reached. At that point, the mechanism is released, and the water pushes the cap out of the way as it discharges water onto the fire (**FIGURE 7-32**).

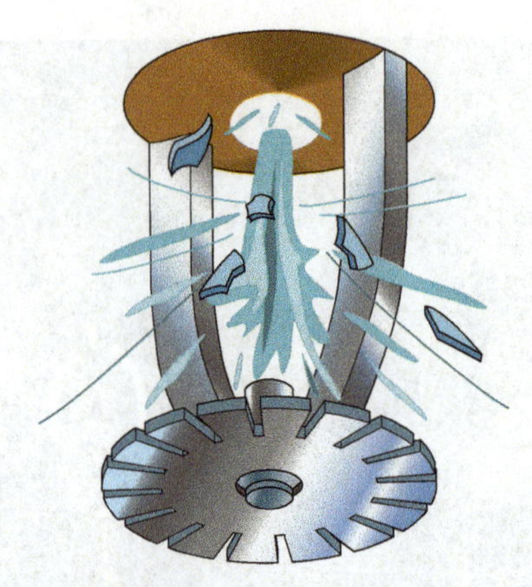

FIGURE 7-32 Automatic sprinkler heads are designed to activate at a wide range of temperatures. How quickly they react depends on their temperature rating.

FIGURE 7-34 Frangible-bulb sprinkler heads activate when the liquid in the bulb expands and breaks the glass.

FIGURE 7-33 In fusible-link sprinkler heads, two pieces of metal are linked together by an alloy such as solder.

FIGURE 7-35 Chemical-pellet sprinkler heads have a plunger mechanism that holds the cap in place.

Fusible-link sprinkler heads use a metal alloy, known as solder, that melts at a specific temperature (**FIGURE 7-33**). This alloy links two pieces of metal that keep the cap in place. When the designated operating temperature is reached, the solder melts and the link breaks, releasing the cap. Fusible-link sprinkler heads come in a wide range of styles and temperature ratings.

Frangible-bulb sprinkler heads use a glass bulb filled with glycerin or alcohol to hold the cap in place (**FIGURE 7-34**). The bulb also contains a small air bubble. As the bulb is heated, the liquid absorbs the air bubble and expands until it breaks the glass, releasing the cap. The volume and composition of the liquid and the size of the air bubble determine the temperature at which the head activates as well as the speed with which it responds.

Chemical-pellet sprinkler heads use a plunger mechanism and a small chemical pellet to hold the cap in place (**FIGURE 7-35**). The chemical pellet will liquefy

PRO TIPS

In automatic sprinklers manufactured before the mid-1950s, the deflectors in both pendant and upright sprinkler heads directed part of the water stream up toward the ceiling. It was believed that this action helped cool the area and extinguish the fire. Sprinkler heads with this design are called old-style sprinklers, and many of them remain in service today.

Automatic sprinklers manufactured after the mid-1950s deflect the entire water stream down toward the fire. These types of heads are referred to as new-style heads or standard spray heads. New-style heads can replace old-style heads, but the reverse is not true. Due to different coverage patterns, old-style heads should not be used to replace any new-style heads.

TABLE 7-2 Temperature Rating Coded by Color of Sprinkler Head

Maximum Ceiling Temperature (°F)	Temperature Rating (°F)	Color Code	Glass Bulb Colors
100	135 to 170	Uncolored or black	Orange or red
150	175 to 225	White	Yellow or green
225	250 to 300	Blue	Blue
300	325 to 375	Red	Purple
375	400 to 475	Green	Black
475	500 to 575	Orange	Black
625	650	Orange	Black

when the temperature reaches a preset point. When the pellet melts, the liquid compresses the plunger, releasing the cap and allowing water to flow.

A typical rating for sprinkler heads in buildings such as an office building is 165°F (73.9°C). The actual temperature of the head is determined by it location. For example, heads in an office might be 165°F (73.9°C), but heads in an attic space, where ambient temperatures are higher, might be 212°F (100°C). If heads are located near space heaters in a parking garage, it would not be surprising to see some set at 285°F (140.6°C). Sprinkler heads use a color-coding system in the liquid to identify the temperature rating (**TABLE 7-2**).

The temperature rating on a sprinkler head must match the anticipated ambient air temperatures. If the rating is too low for the ambient air temperature, accidental activations may occur, which would cause considerable water damage. Conversely, if the rating is too high, the system will be slow to react to a fire, and the fire may be able to establish itself and grow before the system activates.

Standpipe Systems

A **standpipe system** consists of a network of inlets, pipes, and outlets for fire hoses that are built into a structure to provide water for firefighting purposes. It includes one or more inlets supplied by a municipal water supply or fire department hoses using the fire department connection, piping to carry the water closer to the fire, and one or more outlets equipped with valves to which fire hoses can be connected (**FIGURE 7-36**). Standpipe systems are required in many high-rise buildings, and they are found in many other structures as well. Standpipes are also installed to carry water to large bridges and to supply water to limited-access highways that are not equipped with fire hydrants.

FIGURE 7-36 Standpipe outlets allow fire hoses to be connected inside a building.

FIGURE 7-37 A Class I standpipe provides water for fire department hose lines.

Standpipes providing access to the water supply are found in buildings both with and without sprinkler systems. In many newer buildings, sprinklers and standpipes are combined into a single system. In older buildings, the sprinkler and standpipe systems are typically separate entities. Three categories of standpipes—Class I, Class II, and Class III—are distinguished based on their intended use.

Class I Standpipe

A **Class I standpipe** is designed for use by fire department personnel only. Each outlet has a 2½ in. (63.5 mm) male coupling and a valve to open the water supply after the attack line is connected (**FIGURE 7-37**). Responding firefighters carry a fire hose, fire nozzle, and other equipment into the building with them, usually in some sort of roll, bag, or backpack. A Class I standpipe system must be able to supply an adequate volume of water with sufficient pressure to operate fire department attack hose lines.

Class II Standpipes

A **Class II standpipe** is designed for use by the building occupants. The outlets are generally equipped with a length of 1½ in. (38.1 mm) single-jacket hose preconnected to the system (**FIGURE 7-38**). These systems are intended to enable occupants to attack a fire before the fire department arrives, but their safety and effectiveness are questionable. After all, most building occupants are not trained to attack fires safely. If a fire cannot be controlled with a regular fire extinguisher, it is usually safer for the occupants to simply evacuate the building and call the fire department. Class II standpipes may be useful at facilities such as refineries and military bases, where workers are trained as an in-house fire brigade.

FIGURE 7-38 A Class II standpipe is intended to be used by building occupants to attack incipient-stage fires.

Class III Standpipes

A **Class III standpipe** has the features of both Class I and Class II standpipes in a single system. This kind of system has 2½ in. (63.5 mm) outlets for fire department use as well as smaller outlets with attached hoses for occupant use. As in Class II systems, the occupant hoses may have been removed—either intentionally or by vandalism—so the system may basically become a Class I system.

PRO TIPS

On February 23, 1991 a fire occurred on the 22nd floor of the Meridian Plaza high rise office building in Philadelphia, Pennsylvania. One of the many factors that contributed to this out-of-control fire was the incorrect installation of the pressure-reducing hose valves.

FIGURE 7-39 A pressure-reducing valve on a standpipe outlet may be necessary on lower floors to avoid problems caused by backpressure.

Courtesy of Dixon Valve and Coupling.

Water Flow in Standpipe Systems

Standpipes are designed to deliver a minimum amount of water at a particular pressure to each floor. The design requirements depend on the code requirements in effect when the building was constructed. The actual flow also depends on the water supply, as well as on the condition of the piping system and fire pumps. NFPA 14, *Standard for the Installation of Standpipes and Hose Systems*, establishes the design and installation criteria for all standpipe systems.

Pressure-reducing valves (**FIGURE 7-39**) are often installed at the outlets to limit the water pressure. A vertical column of water, such as the water in a standpipe riser, exerts a backpressure (also called head pressure). In a tall building, this backpressure can amount to hundreds of psi (kPa) at lower floor levels. If a hose line is connected to an outlet without a flow restrictor or a pressure-reducing valve, the water pressure could rupture the hose, and the excessive nozzle pressure could make the line difficult or dangerous to handle. Building and fire codes limit the height of a single riser and may also require the installation of pressure-reducing valves on lower floors.

Installation and maintenance of these devices is critical, as these devices can cause problems for firefighters. An improperly adjusted pressure-reducing valve, for example, could severely restrict the flow to a hose line. One cause of this is when the wrong pressure-reducing valve has been installed on a floor.

Water Supplies

Both standpipe systems and sprinkler systems are supplied with water in essentially the same way. That is, many wet standpipe systems in modern buildings are connected to a public water supply, often with an electric or diesel fire pump to provide additional pressure when necessary. Some of these systems also have a water storage tank that serves as a backup supply. In these systems, the FDC on the outside of the building can be used to increase the flow, boost the pressure, or obtain water from an alternative source.

Dry standpipe systems are found in many older buildings. If freezing weather is a problem, such as in open parking structures, bridges, and tunnels, dry standpipe systems are still acceptable. Most dry standpipe systems do not have a permanent connection to a water supply, so the FDC must be used to pump water into the system. If a fire occurs in a building with dry standpipes, connecting the hose lines to the FDC and charging the system with water are high priorities. When a building is being remodeled, code officials might seek the installation of a dry standpipe, which is better than having no standpipe at all.

Some dry standpipe systems are connected to a water supply through a dry-pipe or deluge valve, similar to a sprinkler system. In such systems, opening an outlet valve or tripping a switch next to the outlet releases water into the standpipes. High-rise buildings often incorporate complex systems of risers, storage tanks, and fire pumps at various floors to deliver the needed flows to upper floors.

Documentation and Reporting of Issues

Documentation is crucial as you are stating that the system meets the operational readiness objectives. While most contractors will have forms that they will ask you to sign off on, it is still a good idea that you have your own paperwork detailing the results of the inspection (**FIGURE 7-40**). Request a copy of the contractor's testing forms.

If there are any deficiencies noted in the system, they should be documented and checked again until compliant. It is also important that copies of any testing documentation should be given, at a minimum, to the contractor, the building owner, the municipality, and of course, a copy for the occupancy files for historical purposes.

Inspection Checklist
Sprinkler Systems

Building: ______

Address: ______

Inspector: ______ Date: ______

Date of Last Inspection: ______ **Outstanding Violations:** ❑ Yes ❑ No

General

Date sprinklers installed: ______

Were building alterations/renovations made since last inspection?	❑ Yes	❑ No	
Was new sprinkler system added since last inspection?	❑ Yes	❑ No	
Any sprinkler system alteration made since last inspection?	❑ Yes	❑ No	
What is system type?			
❑ Wet			
❑ Dry			
❑ Preaction			
❑ Deluge			
Is building fully protected with sprinklers?	❑ Yes	❑ No	
If not, explain: ______			
Sprinkler Valves			
Do sprinkler valves appear in good working order?	❑ Yes	❑ No	
Is dry pipe valve in heated enclosure?	❑ Yes	❑ No	❑ N/A*
Are spare sprinklers provided?	❑ Yes	❑ No	
Control Valves			
Are control valves sealed?	❑ Yes	❑ No	❑ N/A
Are they locked?	❑ Yes	❑ No	❑ N/A
Do they have tamper switches?	❑ Yes	❑ No	❑ N/A
Fire Department Connections			
Are fire department connections clear and unobstructed?	❑ Yes	❑ No	❑ N/A
Are protective caps in place?	❑ Yes	❑ No	❑ N/A
Are connections identified?	❑ Yes	❑ No	❑ N/A
Quarterly Inspections and Tests Recorded			
Are quarterly inspections and tests recorded?	❑ Yes	❑ No	

**N/A (not applicable) means there's no such feature in the building.*

FIGURE 7-40 A sample sprinkler system inspection form.

WRAP-UP

CHAPTER SUMMARY

- Fire suppression systems include automatic sprinkler systems, standpipe systems, and specialized extinguishing systems such as dry chemical systems.
- The importance of a dependable and adequate water supply for fire-suppression systems and fire department operations is critical.
- The fire department accesses the water supply via fire hydrants. The two major types of fire hydrants in use today are the dry-barrel hydrant and the wet-barrel hydrant.
- Fire hydrants are located according to local standards and nationally recommended practices. Fire

hydrants may be placed a certain distance apart, perhaps every 500 ft in residential areas and every 300 ft in high-value commercial and industrial areas. Consult your local community codes and standards.

- There are four major classifications of automatic sprinkler systems. Each type of system includes a piping system that contains water from a source of supply to the sprinklers in the area under protection. The four major classifications are:
  - Wet-pipe systems
  - Dry-pipe systems
  - Preaction systems
  - Deluge systems
- Sprinkler heads are the working ends of a sprinkler system. In most systems, the heads serve two functions: activate the sprinkler system and apply water to the fire. Sprinkler heads are composed of a body, or frame, which includes the orifice (opening); a release mechanism, which holds a cap in place over the orifice; and a deflector, which directs the water in a spray pattern. Standard sprinkler heads have a ½ in. (12.7 mm) orifice, but several other sizes are available for special applications.
- A standpipe system consists of a network of inlets, pipes, and outlets for fire hoses that are built into a structure to provide water for firefighting purposes. It includes one or more inlets supplied by a municipal water supply or fire department hoses using the fire department connection, piping to carry the water closer to the fire, and one or more outlets equipped with valves to which fire hoses can be connected. Standpipe systems are required in many high-rise buildings, and they are found in many other structures as well.
- As the first responder inspector, your main objective when performing inspections is to determine the operational readiness of an existing system, that the proper maintenance is being done, and if deficiencies are found that they are reported to the AHJ.

KEY TERMS

Accelerator A device that accelerates the removal of the air from a dry-pipe or preaction sprinkler system.

Alarm valve A valve that signals an alarm when a sprinkler head is activated and prevents nuisance alarms caused by pressure variations.

Automatic sprinkler heads The working ends of a sprinkler system. They serve to activate the system and to apply water to the fire.

Automatic sprinkler system A system of pipes filled with water under pressure that discharges water immediately when a sprinkler head opens.

Butterfly valve A type of indicating valve that moves a piece of metal 90° within the pipe and shows if the water supply is open or closed.

Carbon dioxide extinguishing system A fire suppression system that is designed to protect either a single room or series of rooms by flooding the area with carbon dioxide.

Check valve A valve that allows flow in one direction only. (NFPA 13R)

Chemical-pellet sprinkler head A sprinkler head activated by a chemical pellet that liquefies at a preset temperature.

Class I standpipe A standpipe system designed for use by fire department personnel only. Each outlet should have a valve to control the flow of water and a 2½ in. male coupling for fire hose.

Class II standpipe A standpipe system designed for use by occupants of a building only. Each outlet is generally equipped with a length of 1½ in. single-jacket hose and a nozzle, which are preconnected to the system.

Class III standpipe A combination system that has features of both Class I and Class II standpipes.

Corrosion-resistant sprinklers Sprinkler heads with special coating or plating such as wax or lead to use in potentially corrosive atmospheres.

Deluge sprinkler system A sprinkler system in which all sprinkler heads are open. When an initiation device, such as a smoke detector or heat detector, is activated, the deluge valve opens and water discharges from all of the open sprinkler heads simultaneously.

Deluge valve A valve assembly designed to release water into a sprinkler system when an external initiation device is activated.

Dry chemical extinguishing system An automatic fire extinguishing system that discharges a dry chemical agent.

Dry-barrel hydrant A type of hydrant used in areas subject to freezing weather. The valve that allows water

KEY TERMS CONTINUED

to flow into the hydrant is located underground and the barrel of the hydrant is normally dry.

Dry-pipe sprinkler system A sprinkler system in which the pipes are normally filled with compressed air. When a sprinkler head is activated, it releases the air from the system, which opens a valve so the pipes can fill with water.

Dry-pipe valve The valve assembly on a dry sprinkler system that prevents water from entering the system until the air pressure is released.

Dry sprinkler heads Sprinkler heads that are installed when 40 degrees Fahrenheit cannot be maintained. Dry sprinkler heads are constructed to provide isolation between the head and water supply by use of a cylinder that extends from the head to the threads of the pipe fitting. The threads reside in heated areas or are installed on a dry sprinkler system, and use pen-dent style dry sprinkler heads.

Early-suppression fast-response (ESFR) sprinkler head A sprinkler head designed to react quickly and suppress a fire in its early stages.

Exhauster A device that accelerates the removal of the air from a dry-pipe or preaction sprinkler system.

Fire department connection (FDC) A fire hose connection through which the fire department can pump water into a sprinkler system or standpipe system.

Flow switch An electrical switch that is activated by water moving through a pipe in a sprinkler system.

Flush sprinkler A sprinkler in which all or part of the body, including the shank thread, is mounted above the lower plane of the ceiling. (NFPA 13)

Frangible-bulb sprinkler head A sprinkler head with a liquid-filled bulb. The sprinkler head activates when the liquid is heated and the glass bulb breaks.

Fusible-link sprinkler head A sprinkler head with an activation mechanism that incorporates two pieces of metal held together by low-melting-point solder. When the solder melts, it releases the link and water begins to flow.

Halon 1301 A liquefied gas-extinguishing agent that puts out a fire by chemically interrupting the combustion reaction between fuel and oxygen. Halon agents leave no residue.

Intermediate level sprinklers Sprinklers equipped with integral shields to protect their operating elements from the discharge of sprinklers installed at higher elevations.

Large-drop sprinkler head A sprinkler head that generates large drops of water of such size and velocity as to enable effective penetration of a high-velocity fire plume.

Nozzle sprinkler head Sprinkler heads used in applications requiring special discharge patterns, such as directional spray or fine spray.

Open sprinkler heads A sprinkler that does not have actuators or heat-responsive elements. (NFPA 13)

Ornamental sprinklers Sprinkler that have been painted or plated by the manufacturer.

Outside stem and yoke (OS&Y) valve A sprinkler control valve with a valve stem that moves in and out as the valve is opened or closed.

Pendant sprinkler head A sprinkler head designed to be mounted on the underside of sprinkler piping so that the water stream is directed down.

Post indicator valve (PIV) A sprinkler control valve with an indicator that reads either open or shut depending on its position.

Preaction sprinkler system A dry sprinkler system that uses a deluge valve instead of a dry-pipe valve and requires activation of a secondary device before the pipes will fill with water.

Recessed sprinkler A sprinkler in which all or part of the body, other than the shank thread, is mounted within a recessed housing. (NFPA 13)

Residential sprinkler system A sprinkler system designed to protect dwelling units.

Riser The vertical supply pipes in a sprinkler system. (NFPA 13)

Sidewall sprinklers A sprinkler that is mounted on a wall and discharges water horizontally into a room.

Sprinkler piping The network of piping in a sprinkler system that delivers water to the sprinkler heads.

Standpipe system A system of pipes and hose outlet valves used to deliver water to various parts of a building for fighting fires.

Tamper switch A switch on a sprinkler valve that transmits a signal to the fire alarm control panel if the normal position of the valve is changed.

Upright sprinkler head A sprinkler head designed to be installed on top of the supply piping; it is usually marked SSU ("standard spray upright").

Wall post indicator valve (WPIV) A sprinkler control valve that is mounted on the outside wall of a building. The position of the indicator tells whether the valve is open or shut.

Water-motor gong An audible alarm notification device that is powered by water moving through the sprinkler system.

Wet chemical extinguishing systems An extinguishing system that discharges a proprietary liquid extinguishing agent.

Wet-barrel hydrant A hydrant used in areas that are not susceptible to freezing. The barrel of the hydrant is normally filled with water.

Wet-pipe sprinkler system A sprinkler system in which the pipes are normally filled with water.

You Are the First Responder Inspector

You are inspecting a large industrial complex. The complex owner is new and does not know much about fire equipment. You are pretty much on your own as the person accompanying you is of little help in explaining the fire suppression systems installed in his new complex.

1. The large industrial complex has computer rooms, manufacturing areas, storage areas, large exterior high voltage electrical transformers filled with PCBs, office space, and a cafeteria. What type of sprinkler should you expect to find in this complex?
   A. Wet-pipe systems
   B. Deluge systems
   C. Special types
   D. All of the above
2. The owner is very proud of the servers in his new computer room. What type of fire suppression system should you expect to find in this room?
   A. Deluge system
   B. Dry-pipe system
   C. Clean agent extinguishing system
   D. Combined dry-pipe and preaction system
3. In the manufacturing area, some corrosive materials are used during the production process. When you look up, what type of sprinkler head should you expect to see?
   A. Upright sprinkler head
   B. Corrosion-resistant sprinkler head
   C. Ornamental sprinklers
   D. Intermediate level sprinkler heads
4. After inspecting the interior of the complex, you are walking the grounds to ensure that the complex has a sufficient water supply. You should find the fire hydrants ________ ft apart.
   A. 300
   B. 500
   C. 200
   D. 450

CHAPTER

8

Portable Fire Extinguishers

NFPA 1030 THAT INCLUDES CHAPTER 6: FIRST RESPONDER INSPECTOR (NFPA 1031)

First Responder Inspector

- 6.5.8

ADDITIONAL NFPA STANDARDS

- **NFPA 10**, *Standard for Portable Fire Extinguishers*

KNOWLEDGE OBJECTIVES

After studying this chapter, you will be able to:

1. Define Class A fires.
2. Define Class B fires.
3. Define Class C fires.
4. Define Class D fires.
5. Define Class K fires.
6. Explain the classification and rating system for fire extinguishers.
7. Describe the types of agents used in fire extinguishers.
8. Describe the types of operating systems in fire extinguishers.
9. Describe the basic steps of fire extinguisher operation.
10. Describe how to test the readiness of portable extinguishers in an occupancy.

SKILLS OBJECTIVES

After studying this chapter, you will be able to:

1. Assess the operational readiness of a portable fire extinguisher.
2. Identify the portable fire extinguisher is correctly placed.
3. Confirm the portable fire extinguisher is suited for the hazard.

You Are the First Responder Inspector

You are inspecting a large home improvement store. Before going inside to inspect the interior, you walk around the entire exterior of the building. In the back of the building, you find large pile of mulch next to a loading dock. You note how close this potential fuel is stored next to the building and begin taking a picture for the inspection report. As you are taking the picture, the owner of the building proudly points out the fire extinguisher that is installed in the loading dock bay.

1. Would it be safe to attack a small mulch fire with a portable fire extinguisher?
2. What type of portable fire extinguisher should be installed?

Introduction

Portable fire extinguishers are required in many types of occupancies. Fire extinguishers are used successfully to put out hundreds of fires every day, preventing millions of dollars in property damage as well as saving lives. Most fire extinguishers are easy to operate and can be used effectively by an individual with only basic training.

Fire extinguishers range in size from models that can be operated with one hand to large, wheeled models that contain several hundred pounds of **extinguishing agent** (material used to stop the combustion process) (**FIGURE 8-1**). Extinguishing agents include water, water with different additives, dry-chemicals, wet-chemicals, dry-powders, and gaseous agents. Each agent is suitable for specific types of fires.

Fire extinguishers are designed for different purposes and involve different operational methods. As a first responder inspector, you must know which is the most appropriate kind of extinguisher to use for different types of fires and which kinds must not be used for certain fires. It is your role to ensure that the fire extinguisher is ready to be used in case of a fire.

A.

B.

FIGURE 8-1 Portable fire extinguishers can be large or small. **A.** A wheeled extinguisher. **B.** A one-hand fire extinguisher.

Portable Fire Extinguishers

Purposes of Fire Extinguishers

Portable fire extinguishers have one primary use: to extinguish **incipient** fires (those that have not spread beyond the area of origin). They would not typically be thought of as a replacement for a fire suppression system, if one is called for.

Fire extinguishers are placed in many locations so that they will be available for immediate use on small, incipient-stage fires, such as a fire in a wastebasket. A trained individual with a suitable fire extinguisher could easily control this type of fire. As the flames spread beyond the wastebasket to other contents of the room, however, the fire becomes increasingly difficult to control with only a portable fire extinguisher.

Fire extinguishers are also used to control fires where traditional extinguishing methods are not recommended. For example, using water on fires that involve energized electrical equipment increases the risk of electrocution. Applying water to a fire in a computer or electrical control room could cause extensive damage to the electrical equipment. In these cases, it would be better to use a fire extinguisher with the appropriate extinguishing agent. Special extinguishing agents are also required for fires that involve flammable liquids, cooking oils, and combustible metals. The appropriate type of fire extinguisher should be made available in areas containing these hazards.

Classes of Fires

It is essential to match the appropriate type of extinguisher to the type of fire. Fires and fire extinguishers are grouped into classes according to their characteristics. Some extinguishing agents work more efficiently than others on certain types of fires. In some cases, selecting the proper extinguishing agent will mean the difference between extinguishing a fire and being unable to control it.

More importantly, in some cases it is actually dangerous to apply the wrong extinguishing agent to a fire. Using a water extinguisher on an electrical fire, for example, can cause an electrical shock as well as a short-circuit in the equipment. Likewise, a water extinguisher should never be used to fight a grease fire. Burning grease is generally hotter than 212°F (100°C), so the water converts to steam, which expands very rapidly. If the water penetrates the surface of the grease, the steam is produced within the grease. As the steam expands, the hot grease erupts like a volcano and splatters over everything and everyone nearby, potentially resulting in burns or injuries to people and spreading the fire.

Class A Fires

Class A fires involve ordinary combustibles such as wood, paper, cloth, rubber, household rubbish, and some plastics (**FIGURE 8-2**). Natural vegetation, such as grass and trees, is also Class A material. Water is the most commonly used extinguishing agent for Class A fires, although several other agents can be used effectively.

FIGURE 8-2 Ordinary combustible materials are included in the definition of Class A fires.

FIGURE 8-3 Class B fires involve flammable liquids and gases.

Class B Fires

Class B fires involve flammable or combustible liquids, such as gasoline, oil, grease, tar, lacquer, oil-based paints, and some plastics (**FIGURE 8-3**). Fires involving flammable gases, such as propane or natural gas, are also categorized as Class B fires. Examples of Class B fires include a fire in a pot of molten roofing tar, a fire involving splashed fuel on a hot lawnmower engine, and burning natural gas that is escaping from a gas meter struck by a vehicle. Several different types of extinguishing agents are approved for use in Class B fires.

Class C Fires

Class C fires involve energized electrical equipment, which includes any device that uses, produces, or delivers electrical energy (**FIGURE 8-4**). A Class C fire could involve building wiring and outlets, fuse boxes, circuit

FIGURE 8-4 Class C fires involve energized electrical equipment or appliances.

© Scott Leman/Shutterstock

FIGURE 8-5 Combustible metals in Class D fires require special extinguishing agents.

© Andrew Lambert Photography/Science Source

breakers, transformers, generators, or electric motors. Power tools, lighting fixtures, household appliances, and electronic devices such as televisions, radios, and computers could be involved in Class C fires as well. The equipment must be plugged in or connected to an electrical source, but not necessarily operating, for the fire to be classified as Class C.

Electricity does not burn, but electrical energy can generate tremendous heat that could ignite nearby Class A or B materials. As long as the equipment is energized, the incident must be treated as a Class C fire. Agents that will not conduct electricity, such as dry-chemicals or carbon dioxide, must be used on Class C fires.

Class D Fires

Class D fires involve combustible metals such as magnesium, titanium, zirconium, sodium, lithium, and potassium. Special extinguishing agents are required to fight combustible metals fires (**FIGURE 8-5**). Normal extinguishing agents can react violently—even explosively—if they come in contact with burning metals. Violent reactions also can occur when water strikes burning combustible metals.

Class D fires are most often encountered in industrial occupancies, such as machine shops and repair shops, as well as in fires involving aircraft and automobiles. Magnesium and titanium—both combustible metals—are used to produce automotive and aircraft parts because they combine high strength with light weight. Sparks from cutting, welding, or grinding operations could ignite a Class D fire, or the metal items could become involved in a fire that originated elsewhere.

FIGURE 8-6 Class K fires involve cooking oils and fats.

© Scott Leman/Shutterstock

Class K Fires

Class K fires involve combustible cooking oils and fats (**FIGURE 8-6**). This is a relatively new classification; cooking oil fires were previously classified as Class B combustible liquid fires. The use of high-efficiency modern cooking equipment and the trend toward using vegetable oils instead of animal fats to fry foods in recent years have resulted in higher cooking temperatures. These higher temperatures required the development of a new class of wet-chemical extinguishing agents. Some restaurants continue to use extinguishing agents that were approved for Class B fires, but these agents are not as effective for the new cooking oil fires, as are Class K extinguishers. If a restaurant has a hood and duct extinguishing system, a Class K extinguisher is required in the area of the cooking line. However, the traditional B type extinguisher may still be present for the other possible fires in the kitchen.

Classification of Fire Extinguishers

Portable fire extinguishers are classified and rated based on their characteristics and capabilities. This information is important for selecting the proper extinguisher to fight a particular fire (**TABLE 8-1**). It is also used to determine which types of fire extinguishers should be placed in a given location so that incipient fires can be quickly controlled.

In the United States, **Underwriters Laboratories, Inc. (UL)** is the organization that developed the standards, classification, and rating system for portable fire extinguishers. The UL system identifies the classes of fires for which each fire extinguisher is both safe and effective.

The classification system for fire extinguishers uses both letters and numbers. The letters indicate the classes of fire for which the extinguisher can be used, and the numbers indicate its effectiveness. Numbers are used to rate an extinguisher's effectiveness only for Class A and Class B fires. Fire extinguishers that are safe and effective for more than one class will be rated with multiple letters. For example, an extinguisher that is safe and effective for Class A fires will be rated with an "A"; one that is safe and effective for Class B fires will be rated with a "B"; and one that is safe and effective for both Class A and Class B fires will be rated with both an "A" and a "B."

TABLE 8-1 Types of Fires

Class A	Ordinary combustibles
Class B	Flammable or combustible liquids
Class C	Energized electrical equipment
Class D	Combustible metals
Class K	Combustible cooking media

The number on the Class A and Class B fire extinguishers indicates the relative effectiveness of the fire extinguisher in the hands of a non-expert user. The higher the number, the greater the extinguishing capability of the extinguisher. To receive a 1-A rating the fire extinguisher must extinguish a certain amount of a wood crib fire. A 2-A rating is able to extinguish twice the fire a 1-A does, a 4-A twice the amount a 2-A does. The effectiveness of Class B extinguishers is based on the approximate area (measured in square feet) of burning fuel they are capable of extinguishing. A 10-B rating indicates that a non-expert user should be able to extinguish a fire in a pan of flammable liquid that is 10 square feet in surface area—that is just over 3 ft × 3 ft (91.3 cm × 91.4 cm), not 2 ft × 5 ft (61.0 cm × 152.4 cm) or 1 ft × 10 ft (30.5 cm × 405.8 cm). An extinguisher rated 40-B should be able to control a flammable liquid pan fire with a surface area of 40 square feet (3.72 m^2). An experienced operator should be able to extinguish 2½ times the extinguisher rating. So an experienced operator should be able to take a 40-B extinguisher and extinguish 100 square feet (9.29 m^2) of fire.

There are no numerical ratings for Class-C fires. Class-C simply means the agent will not conduct electricity. If the fire extinguisher can also be used for Class C fires, it contains an agent proven to be nonconductive to electricity and safe for use on energized electrical equipment. For instance, a fire extinguisher that carries a 2-A:10-B:C rating can be used on Class A, Class B, and Class C fires. It has the extinguishing capabilities of a 2-A extinguisher when applied to Class A fires, has the capabilities of a 10-B extinguisher when applied to Class B fires, and can be used safely on energized electrical equipment.

A Class K extinguisher would not be good to use on live electrical equipment due to the chance of the agent conducting electricity, or on Class D fires for to an explosion possibility. In addition to the Class K rating, many extinguishers may also be rated for Class A fires. Their classification could then be 2A:K.

Standard test fires are used to rate the effectiveness of fire extinguishers. This testing may involve different agents, amounts, application rates, and application methods. Fire extinguishers are rated for their ability to control a specific type of fire as well as for the extinguishing agent's ability to prevent rekindling. Some agents can successfully suppress a fire, but are unable to prevent the material from reigniting. A rating is given only if the extinguisher completely extinguishes the standard test fire and prevents rekindling.

Labeling of Fire Extinguishers

Fire extinguishers that have been tested and approved by an independent laboratory are labeled to clearly designate the classes of fire the unit is capable of extinguishing safely. This traditional lettering system has been used for many years and is still found on many fire extinguishers. More recently, a universal pictograph system, which does not require the user to be familiar with the alphabetic codes for the different classes of fires, has been developed.

Traditional Lettering System

The traditional lettering system uses the following labels (**FIGURE 8-7**):

- Extinguishers suitable for use on Class A fires are identified by the letter "A" on a solid green triangle. The triangle has a graphic relationship to the letter "A."
- Extinguishers suitable for use on Class B fires are identified by the letter "B" on a solid red square. Again, the shape of the letter mirrors the graphic shape of the box.
- Extinguishers suitable for use on Class C fires are identified by the letter "C" on a solid blue circle, which also incorporates a graphic relationship between the letter "C" and the circle.
- Extinguishers suitable for use on Class D fires are identified by the letter "D" on a solid yellow, five-pointed star.
- Extinguishers suitable for use on Class K (combustible cooking oil) fires are identified by a pictograph showing a fire in a frying pan.

Pictograph Labeling System

The pictograph system, such as described for Class K fire extinguishers, uses symbols rather than letters on the labels. This system also clearly indicates whether an extinguisher is inappropriate for use on a particular class of fire. The pictographs are all square icons, each of which is designed to represent a certain class of fire (**FIGURE 8-8**). The icon for Class A fires is a burning trashcan beside a wood fire. The Class B fire extinguisher icon is a flame and a gasoline can; the Class C icon is a flame and an electrical plug and socket. The tanks for Class A, B, and C extinguishers are usually the color red. The icon for Class D extinguishers depicts a metal gear, and the tank is usually colored yellow. Extinguishers rated for fighting Class K fires are labeled with an icon showing a fire in a frying pan and are usually chrome.

Under this pictograph labeling system, the presence of an icon indicates that the extinguisher has been rated for that class of fire. A missing icon indicates that the extinguisher has not been rated for that class of fire. A red slash across an icon indicates that the extinguisher must not be used on that type of fire, because doing so would create additional risk.

An extinguisher rated for Class A fires only would show all three icons, but the icons for Class B and Class C would have a red diagonal line through them. This three-icon array signifies that the extinguisher contains a water-based extinguishing agent, making it unsafe to use on flammable liquid or electrical fires.

Certain extinguishers labeled as appropriate for Class B and Class C fires do not include the Class A icon, but may be used to put out small Class A fires. The fact that they have not been rated for Class A fires

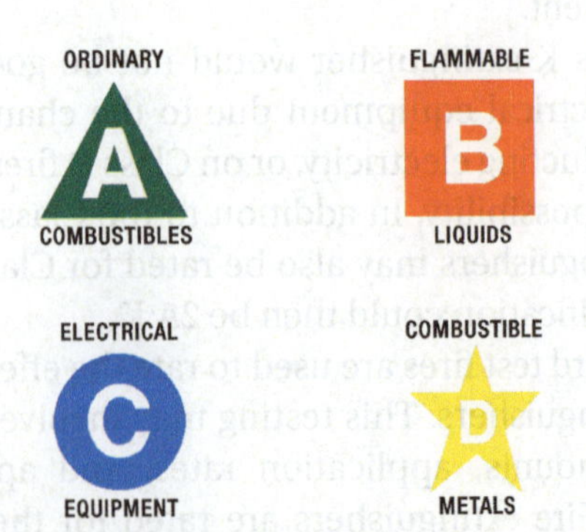

FIGURE 8-7 Traditional letter labels on fire extinguishers often incorporated a shape as well as a letter.

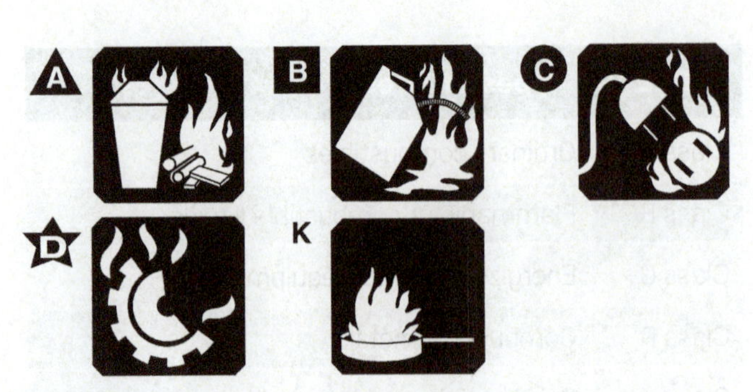

FIGURE 8-8 The icons for Classes A, B, C, D, and K fires.

indicates that they are less effective in extinguishing a common combustible fire than a comparable Class A extinguisher would be.

Fire Extinguisher Placement

Fire codes and regulations require the installation of fire extinguishers in many areas so that they will be available to fight incipient fires. NFPA 10, *Standard for Portable Fire Extinguishers,* lists the requirements for placing and mounting portable fire extinguishers, the appropriate mounting heights, and ongoing testing and maintenance of extinguishers.

The regulations for each type of occupancy specify the maximum floor area that can be protected by each extinguisher, the maximum travel distance from the closest extinguisher to a potential fire, and the types of fire extinguishers that should be provided. Two key factors must be considered when determining which type of extinguisher should be placed in each area: the class of fire that is likely to occur and the potential magnitude of an incipient fire.

Extinguishers must be mounted so they are readily visible and easily accessible (**FIGURE 8-9**). Heavy extinguishers should not be mounted high on a wall. If the extinguisher is mounted too high, a smaller person might be unable to lift it off its hook or could be injured in the attempt.

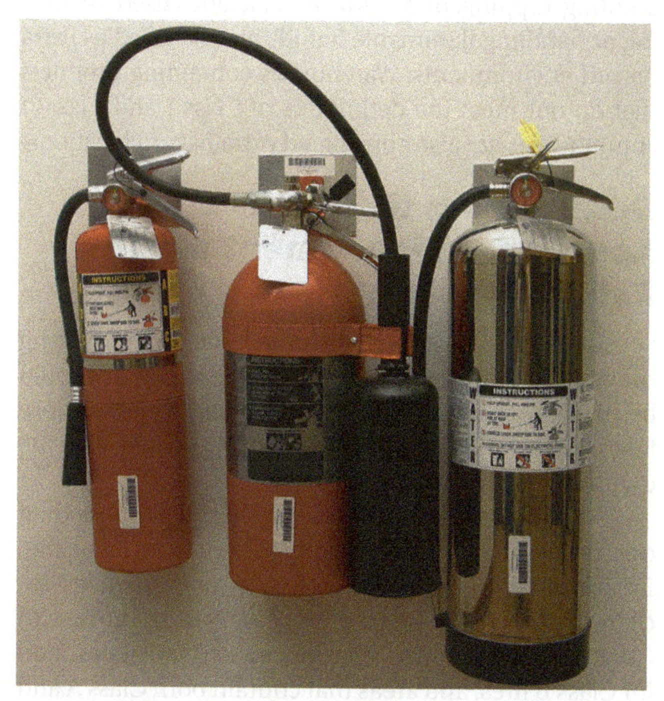

FIGURE 8-9 Extinguishers should be mounted in locations with unobstructed access and visibility.

According to NFPA 10, the required mounting heights for the placement of fire extinguishers are as follows:

- Fire extinguishers weighing up to 40 lb (18.14 kg) should be mounted so that the top of the extinguisher is not more than 5 ft (1.53 m) above the floor.
- Fire extinguishers weighing more than 40 lb (18.14 kg) should be mounted so that the top of the extinguisher is not more than 3.5 ft (1.07 m) above the floor.
- The bottom of an extinguisher should be at least 4 in. (10.2 cm) above the floor.

Classifying Area Hazards

Areas are divided into three risk classifications—light, ordinary, and extra hazard—based on the amount and type of combustibles that are present, including building materials, contents, decorations, and furniture. The quantity of combustible materials present is sometimes called a building's **fire load**. It means the quantity of heat which would be released by the combustion of all the combustible materials. Many objects in a building, such as a large metal cutting press, may weigh a lot, but don't burn, therefore they would not figure into the fire load. The larger the fire load, the larger the potential fire.

The occupancy use category does not necessarily determine the building's hazard classification. The recommended hazard classifications for different types of occupancies are simply guidelines based on typical situations. The hazard classification for each area should be based on the actual amount and type of combustibles that are present.

Light (Low) Hazard

Light (or low) hazard locations are areas where the majority of materials are noncombustible or arranged so that a fire is not likely to spread. Light hazard environments usually contain limited amounts of Class A combustibles, such as wood, paper products, cloth, and similar materials. A light hazard environment might also contain some Class B combustibles (flammable liquids and gases), such as copy machine chemicals or modest quantities of paints and solvents, but all Class B materials must be kept in closed containers and stored safely. Examples of common light hazard environments are most offices, classrooms, churches, assembly halls, and hotel guest rooms (**FIGURE 8-10**).

Ordinary (Moderate) Hazard

Ordinary (or moderate) hazard locations contain more Class A and Class B materials than do light hazard

FIGURE 8-10 Light hazard areas include offices, churches, and classrooms.

FIGURE 8-11 Auto showrooms, hotel laundry rooms, and parking garages are classified as ordinary hazard areas.

locations. Typical examples of ordinary hazard locations include retail stores with on-site storage areas, light manufacturing facilities, auto showrooms, parking garages, research facilities, and workshops or service areas that support light hazard locations, such as hotel laundry rooms or restaurant kitchens (**FIGURE 8-11**).

Ordinary hazard areas also include warehouses that contain Class I and Class II commodities. Class I commodities include noncombustible products stored on wooden pallets or in corrugated cartons that are shrink-wrapped or wrapped in paper. Class II commodities include noncombustible products stored in wooden crates or multilayered corrugated cartons.

Extra (High) Hazard

Extra (or high) hazard locations contain more Class A combustibles and/or Class B flammables than do ordinary hazard locations. Typical examples of extra hazard areas include woodworking shops; service or repair

FIGURE 8-12 Kitchens, woodworking shops, and auto repair shops are considered possible extra hazard locations.

facilities for cars, aircraft, or boats; and many kitchens and other cooking areas that have deep fryers, flammable liquids, or gases under pressure (**FIGURE 8-12**). In addition, areas used for manufacturing processes such as painting, dipping, or coating, and facilities used for storing or handling flammable liquids are classified as extra hazard environments. Warehouses containing products that do not meet the definitions of Class I and Class II commodities are also considered extra hazard locations.

Determining the Most Appropriate Placement of Fire Extinguishers

Several factors must be considered when determining the number and types of fire extinguishers that should be placed in each area of an occupancy. Among these factors are the types of fuels found in the area and the quantities of those materials.

Some areas may need extinguishers with more than one rating or more than one type of fire extinguisher. Environments that include Class A combustibles require an extinguisher rated for Class A fires; those with Class B combustibles require an extinguisher rated for Class B fires; and areas that contain both Class A and Class B combustibles require either an extinguisher that is rated for both types of fires or a separate extinguisher for each class of fire.

Most building or fire codes require extinguishers that are suitable for fighting Class A fires because ordinary combustible materials—such as furniture, partitions, interior finish materials, paper, and packaging products—are so common. Even where other classes of products are used or stored, there is still a need to defend the facility from a fire involving common combustibles.

A single multipurpose extinguisher is generally less expensive than two individual fire extinguishers and eliminates the problem of selecting the proper extinguisher for a particular fire. However, it is sometimes more appropriate to install Class A extinguishers in general-use areas and to place extinguishers that are especially effective in fighting Class B or Class C fires near those specific hazards.

Some facilities present a variety of conditions. In these occupancies, each area must be individually evaluated so that extinguisher installation is tailored to the particular circumstances. A restaurant is a good example of this situation. The dining areas contain common combustibles, such as furniture, tablecloths, and paper products, that would require an extinguisher rated for Class A fires. In the restaurant's kitchen, where the risk of fire involves cooking oils, a Class K extinguisher would provide the best defense.

Similarly, within a hospital, extinguishers for Class A fires would be appropriate in hallways, offices, lobbies, and patient rooms. Class B extinguishers should be mounted in laboratories and areas where flammable anesthetics are stored or handled. Electrical rooms should have extinguishers that are approved for use on Class C fires, whereas hospital kitchens would need Class K extinguishers.

Types of Extinguishing Agents

An extinguishing agent is the substance contained in a portable fire extinguisher that puts out a fire. A variety of chemicals, including water, are used in portable fire extinguishers. The best extinguishing agent for a particular hazard depends on several factors, including the types of materials involved and the anticipated size of the fire. Portable fire extinguishers use seven basic types of extinguishing agents:

- Water
- Dry-chemicals
- Carbon dioxide
- Foam
- Wet-chemicals
- Halogenated agents
- Dry-powder

Water

Water is an efficient, plentiful, and inexpensive extinguishing agent. When it is applied to a fire, it quickly converts from liquid into steam, absorbing great quantities of heat in the process. As the heat is removed from the combustion process, the fuel cools below its ignition temperature and the fire stops burning.

Water is an excellent extinguishing agent for Class A fires. Many Class A fuels will absorb liquid water, which further lowers the temperature of the fuel. This also prevents rekindling.

Water is a much less effective extinguishing agent for other classes of fires. Applying water to hot cooking oil, for example, can cause splattering, which can spread the fire and possibly endanger the operator of the fire extinguisher. Many burning flammable liquids will simply float on top of water. Because water conducts electricity, it is dangerous to apply a stream of water to any fire that involves energized electrical equipment. If water is applied to a burning combustible metal, a violent reaction can occur. Because of these limitations, plain water is used only in Class A fire extinguishers.

One notable disadvantage of water is that it freezes at 32°F (0°C). In areas that are subject to below-freezing temperatures, **loaded-stream fire extinguishers** can be used to counteract this limitation (**FIGURE 8-13**).

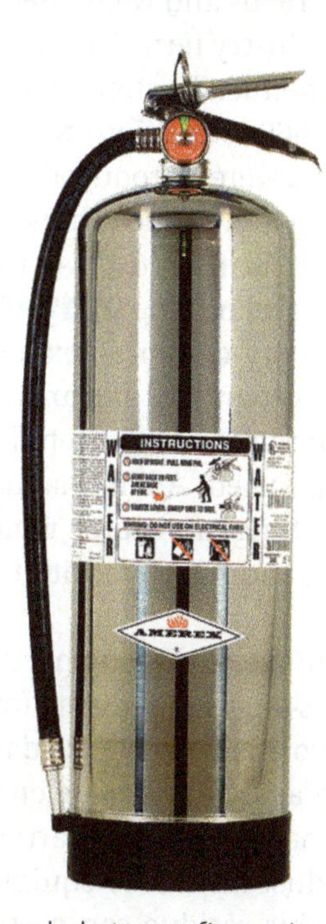

FIGURE 8-13 Loaded-stream fire extinguisher.

Courtesy of Amerex Corporation.

These extinguishers combine an alkali metal salt with water. The salt lowers the freezing point of water, so the extinguisher can be used in much colder areas.

Wetting agents can also be added to the water in a fire extinguisher. These agents reduce the surface tension of the water, allowing it to penetrate more effectively into many fuels, such as baled cotton or fibrous materials.

Dry Chemicals

Dry-chemical fire extinguishers deliver a stream of very finely ground particles onto a fire. Different chemical compounds are used to produce extinguishers of varying capabilities and characteristics. The dry-chemical extinguishing agents works to interrupt the chemical chain reactions that occur as part of the combustion process.

Dry-chemical extinguishing agents offer several advantages over water extinguishers:

- They are effective on Class B (flammable liquids and gases) fires.
- They can be used on Class C (energized electrical equipment) fires, because the chemicals are nonconductive.
- They are not subject to freezing.
- They are smaller and less expensive.

The first dry-chemical extinguishers were introduced during the 1950s and were rated only for Class B and C fires. The industry term for these B:C-rated units is "ordinary dry-chemical" extinguishers.

During the 1960s, **multipurpose dry-chemical fire extinguishers** were introduced. These extinguishers are rated for Class A, B, and C fires. The chemicals in these extinguishers form a crust over Class A combustible fuels to prevent rekindling (**FIGURE 8-14**).

FIGURE 8-14 Multipurpose dry-chemical extinguishers can be used for Class A, B, and C fires.

Multipurpose dry-chemical extinguishing agents take the form of fine particles and are treated with other chemicals to help maintain an even flow when the extinguisher is being used. Additional additives prevent them from absorbing moisture, which could cause packing or caking and interfere with the extinguisher's discharge.

One disadvantage of dry-chemical extinguishers is that the chemicals—particularly the multipurpose dry-chemicals—are corrosive and can damage electronic equipment, such as computers, telephones, and copy machines. The fine particles are carried by the air and settle like a fine dust inside the equipment. Over a period of months, this residue can corrode metal parts, causing considerable damage. If electronic equipment is exposed to multipurpose dry-chemical extinguishing agents, it should be cleaned professionally within 48 hours after exposure.

Five compounds are used as the primary dry-chemical extinguishing agents:

- Sodium bicarbonate (rated for Class B and C fires only)
- Potassium bicarbonate (rated for Class B and C fires only)
- Urea-based potassium bicarbonate (rated for Class B and C fires only)
- Potassium chloride (rated for Class B and C fires only)
- Ammonium phosphate (rated for Class A, B, and C fires)

Sodium bicarbonate is often used in small household extinguishers. Potassium bicarbonate, potassium chloride, and urea-based potassium bicarbonate all have greater fire-extinguishing capabilities (per unit volume) for Class B fires than does sodium bicarbonate. Potassium chloride is more corrosive than the other dry-chemical extinguishing agents.

Ammonium phosphate is the only dry-chemical extinguishing agent that is rated as suitable for use on Class A fires. Although ordinary dry-chemical extinguishers can also be used against Class A fires, a water dousing is also needed to extinguish any smoldering embers and prevent rekindling.

The selection of which dry-chemical extinguisher to use depends on the compatibility of different agents with one another and with any products that they might contact. Some dry-chemical extinguishing agents cannot be used in combination with particular types of foam.

Carbon Dioxide

Carbon dioxide (CO_2) is a gas that is 1.5 times heavier than air. When carbon dioxide is discharged on a fire, it forms a dense cloud that displaces the air surrounding the fuel. This effect interrupts the combustion process by reducing the amount of oxygen that can reach the fuel. A blanket of carbon dioxide over the surface of a liquid fuel can also disrupt the fuel's ability to vaporize.

In portable **carbon dioxide fire extinguishers**, carbon dioxide is stored under pressure as a colorless and odorless liquid. It is discharged through a hose and expelled on the fire through a horn. When it is released, the carbon dioxide is very cold and forms a visible cloud of "dry ice" when moisture in the air freezes as it comes into contact with the carbon dioxide.

Carbon dioxide is rated for Class B and C fires only. This extinguishing agent does not conduct electricity and has two significant advantages over dry-chemical agents: It is not corrosive and it does not leave any residue.

Carbon dioxide also has several limitations and disadvantages:

- Weight: Carbon dioxide extinguishers are heavier than similarly rated extinguishers that use other extinguishing agents (**FIGURE 8-15**).
- The ratings on carbon dioxide extinguishers are far less than a typical multi-purpose fire extinguisher.
- Range: Carbon dioxide extinguishers have a short discharge range, which requires the operator to be close to the fire, increasing the risk of personal injury.
- Weather: Carbon dioxide does not perform well at temperatures below 0°F (-18°C) or in windy or drafty conditions, because it dissipates before it reaches the fire.
- Confined spaces: When used in confined areas, carbon dioxide dilutes the oxygen in the air. If it is diluted enough, people in the space can begin to suffocate.
- Suitability: Carbon dioxide extinguishers are not suitable for use on fires involving pressurized fuel or on cooking grease fires.

FIGURE 8-15 Carbon dioxide extinguishers are heavy owing to the weight of the container and the quantity of agent needed to extinguish a fire. They also have a large discharge nozzle, making them easily identifiable.

Foam

Foam fire extinguishers discharge a water-based solution with a measured amount of foam concentrate added. The nozzles on foam extinguishers are designed to introduce air into the discharge stream, thereby producing a foam blanket. Foam extinguishing agents are formulated for use on either Class A or Class B fires.

Class A foam extinguishers for ordinary combustible fires extinguish fires in the same way that water extinguishes fires. This type of extinguisher can be produced by adding Class A foam concentrate to the water in a standard 2.5-gallon (9.5-L), stored-pressure extinguisher. The foam concentrate reduces the surface tension of the water, allowing for its better penetration into the burning materials.

Class B foam extinguishers discharge a foam solution that floats across the surface of a burning liquid and prevents the fuel from vaporizing. This foam blanket forms a barrier between the fuel and the oxygen, extinguishing the flames and preventing reignition. These agents are not suitable for Class B fires that involve pressurized fuels or cooking oils.

The most common Class B additives are **aqueous film-forming foam (AFFF)** and **film-forming fluoroprotein (FFFP) foam** (**FIGURE 8-16**). Both concentrates produce very effective foams. Which one

FIGURE 8-16 An AFFF extinguisher produces an effective foam for use on Class B fires.

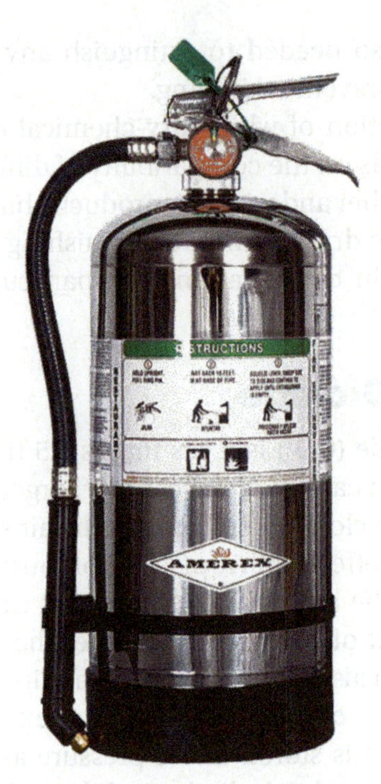

FIGURE 8-17 Wet-chemical fire extinguisher.

should be used depends on the product's compatibility with a particular flammable liquid and other extinguishing agents that could be used on the same fire.

Some Class B foam extinguishing agents are approved for use on **polar solvents**—that is, water-soluble flammable liquids such as alcohols, acetone, esters, and ketones. Only extinguishers that are specifically labeled for use with polar solvents should be used if these products are present.

Although they are not specifically intended to extinguish Class A fires, most Class B foams can also be used on ordinary combustibles. The reverse is not true, however: Class A foams are not effective on Class B fires. Foam extinguishers are not suitable for use on Class C fires and cannot be stored or used at freezing temperatures.

Wet Chemicals

Wet-chemical fire extinguishers are the only type of extinguisher to qualify under the Class K rating requirements (**FIGURE 8-17**). They use **wet-chemical extinguishing agents**, which are chemicals applied as water solutions. Before Class K extinguishing agents were developed, most fire-extinguishing systems for kitchens used dry-chemicals. The minimum requirement in a commercial kitchen was a 40-B-rated sodium bicarbonate or potassium bicarbonate extinguisher. These systems required extensive clean-up after their use, which often resulted in serious business interruptions.

All new fixed extinguishing systems in restaurants and commercial kitchens now use wet-chemical extinguishing agents. These agents are specifically formulated for use in commercial kitchens and food-product manufacturing facilities, especially where food is cooked in a deep fryer. The fixed systems discharge the agent directly over the cooking surfaces. There is no numeric rating of their efficiency in portable fire extinguishers. The Class K wet-chemical agents include aqueous solutions of potassium acetate, potassium carbonate, and potassium citrate, either singly or in various combinations. These wet agents convert the fatty acids in cooking oils or fats to a soap or foam, a process known as **saponification**.

When wet-chemical agents are applied to burning vegetable oils, they create a thick blanket of foam that quickly smothers the fire and prevents it from reigniting while the hot oil cools. The agents are discharged as a fine spray, which reduces the risk of splattering. They are very effective at extinguishing cooking oil fires, and clean-up afterward is much easier, allowing a business to reopen sooner.

Halogenated Agents

Halogenated extinguishing agents are produced from a family of liquified gases, known as halogens, that includes fluorine, bromine, iodine, and chlorine. Hundreds of different formulations can be produced from these elements; these myriad versions have many different properties and potential uses. Although several of these formulations are very effective for extinguishing fires, only a few of them are commonly used as extinguishing agents. The first member of the halon products was Halon 104, more commonly known as carbon tetrachloride.

Halogenated extinguishing agents are commonly called **clean agents** because they leave no residue and are ideally suited for areas that contain computers or sensitive electronic equipment. Per pound, they are approximately twice as effective at extinguishing fires as is carbon dioxide. The principle behind extinguishing is that clean agents interfere with the combustion process.

Two categories of halogenated extinguishing agents are distinguished: halons and halocarbons. A 1987 international agreement, known as the Montreal Protocol, limited halon production because these agents damage the earth's ozone layer. Halons have since been replaced by a new family of extinguishing agents, halocarbons.

The halogenated agents are stored as liquids and are discharged under relatively high pressure. They release a mist of vapor and liquid droplets that disrupts the molecular chain reactions within the combustion process, thereby extinguishing the fire. These agents dissipate rapidly in windy conditions, as does carbon dioxide, so their effectiveness is limited in outdoor locations. Because halogenated agents also displace oxygen, they should be used with care in confined areas.

Halon agents (bromochlorodifluoromethane) were commonly used many years ago. However, as a result of the Montreal Protocol, the United States banned the production and importation of halon in the mid 1990s. Halon can still be acquired from some companies that recapture the chemical from old systems. Due to the human health concerns, care should be taken around these systems should they be discharged. Currently, four types of halocarbon agents are used in portable extinguishers: hydrochlorofluorocarbon (HCFC), hydrofluoro-carbon (HFC), perfluorocarbon (PFC), and fluoroiodocarbon (FIC). Some of the trade names of replacement products used today are Halatron, FM200, Inergen, ECARO-25, and FE 36.

Dry-Powder

Dry-powder extinguishing agents are chemical compounds used to extinguish fires involving combustible metals (Class D fires). These agents are stored in fine granular or powdered form and are applied to smother the fire. They form a solid crust over the burning metal, which both blocks out oxygen (the fuel for the fire) and absorbs heat.

The most commonly used dry-powder extinguishing agent is formulated from finely ground sodium chloride (table salt) plus additives to help it flow freely over a fire. A thermoplastic material mixed with the agent binds the sodium chloride particles into a solid mass when they come into contact with a burning metal.

Another dry-powder agent is produced from a mixture of finely granulated graphite powder and phosphorus-containing compounds. This agent cannot be expelled from fire extinguishers; instead, it is produced in bulk form and applied by hand, using a scoop or a shovel. When applied to a metal fire, the phosphorus compounds release gases that blanket the fire and cut off its supply of oxygen; the graphite absorbs heat from the fire, allowing the metal to cool below its ignition point. Other specialized dry-powder extinguishing agents are available for fighting specific types of metal fires. For details, see NFPA's *Fire Protection Handbook*.

Class D agents must be applied very carefully so that the molten metal does not splatter. No water should come in contact with the burning metal, as even a trace quantity of moisture can cause a violent reaction.

Fire Extinguisher System Readiness

Fire extinguishers must always be in a state of readiness. A standard checklist may help you in assessing fire extinguishers (**FIGURE 8-18**). As an FRI, you assess the state of readiness by inspecting:

- The travel distances to the fire extinguisher—The travel distance for a Class A fire extinguisher is 75 ft (22.9 m). For a Class B fire extinguisher, the proper travel distance is determined by the hazard being protected. NFPA 10 will dictate the distance and provides a table stating how many extinguisher are needed for a given floor area.
- The mounting of the fire extinguisher—Most fire extinguishers are mounted with the top not higher that 5 ft (1.52 m) above the floor. If the fire extinguisher is more than 40 lb (18.1 kg), then the top must be no higher than 3.5 ft (1.07 m) above the floor. These distances make it easy for an operator to pick up the fire extinguisher. In either case, the bottom of the extinguisher must be a minimum of 4 in. (101.6 mm) off of the floor. This allows for easy cleaning of the floor. The fire extinguisher must

Office of Fire Inspection Portable Fire Extinguisher Checklist

	Code Requirements	Pass/Fail	Required Corrective Action
Travel distance			
Number per floor			
Mounting			
Clear access			
Clear view			
Type			
	Physical Inspection	**Pass/Fail**	**Required Corrective Action**
No visible damage to cylinder			
Hose attached securely			
No obstructions in hose			
Safety pin in place			
Proper gauge pressure			
	Inspection Tag	**Pass/Fail**	**Required Corrective Action**
Current vendor inspection tag			
Tag information matches extinguisher label			

FIGURE 8-18 A sample checklist for fire extinguisher assessment.

be mounted to prevent it from being moved to the back of a shelf or behind a piece of furniture. The fire extinguisher should be mounted near a normal means of egress and by an exit.

- Access to the fire extinguisher—The fire extinguisher should be easily accessible. No item should be blocking the fire extinguisher from view or access.
- The type of fire extinguisher—The proper type of fire extinguisher should be available. It would not be proper to have a water extinguisher in an electric room or a flammable liquids room. To prevent the chance of using the wrong fire extinguisher, most businesses use a multi-purpose extinguisher. These extinguishers are good for the three common classes of fire A, B, and C. An ABC type and a BC type extinguisher look identical, and they only way to tell them apart is to look at the rating or pictographs on the fire extinguisher.

When you inspect fire extinguishers, look for physical damage to the container: ensure that the hose is actually attached, no foreign matter is in the hose, the safety pin is in place, the seal holds the pin in place, and the gauge is showing the proper operating pressure. The location of the fire extinguisher should be considered as well. It should not be placed in an area where it will be subjected to possible damage. This would mean that it should not protrude into a means of egress more than 4 in. (101.6 mm). A fire extinguisher must have current inspection tags proving that an outside company inspected and tagged each unit. The tag is a visual indicator that the fire extinguisher was functional at the time of the outside company's inspection.

PRO TIPS

NFPA 10 mandates that fire extinguishers undergo regular inspection, testing, and maintenance by qualified personnel. This includes an annual inspection to ensure functionality. After inspection, a tag by the technician must be attached to the extinguisher, documenting the date and details. Proper maintenance ensures reliability in emergencies.

Documentation and Reporting of Issues

It is common to find deficiencies during the course of a fire inspection. Examples include that fire extinguisher has no pressure, the fire extinguisher was used, or the hose is not connected to the fire extinguisher. These items must be noted in the fire inspection report. In the inspection report, these items are easily addressed by writing that the fire extinguisher must be in proper working order with a new inspection tag. For issues such as fire extinguishers not being mounted, fire extinguishers that cannot be accessed, or fire extinguishers that are missing, write these as violations and re-inspect for code compliance at a later date. A copy of the fire inspection report should be left with the building owner as a reference to correct the violations.

WRAP-UP

CHAPTER SUMMARY

- Portable fire extinguishers have one primary use: to extinguish incipient fires (those that have not spread beyond the area of origin). They would not typically be thought of as a replacement for a fire suppression system, if one is called for.
- It is essential to match the appropriate type of extinguisher to the type of fire. In some cases it is actually dangerous to apply the wrong extinguishing agent to a fire.
- Class A fires involve ordinary combustibles such as wood, paper, cloth, rubber, household rubbish, and some plastics.
- Class B fires involve flammable or combustible liquids, such as gasoline, oil, grease, tar, lacquer, oil-based paints, and some plastics. Fires involving flammable gases, such as propane or natural gas, are also categorized as Class B fires.
- Class C fires involve energized electrical equipment, which includes any device that uses, produces, or delivers electrical energy. A Class C fire could involve building wiring and outlets, fuse boxes, circuit breakers, transformers, generators, or electric motors.
- Class D fires involve combustible metals such as magnesium, titanium, zirconium, sodium, lithium, and potassium. Special extinguishing agents are required to fight combustible metals fires.
- Class K fires involve combustible cooking oils and fats.
- Portable fire extinguishers are classified and rated based on their characteristics and capabilities:
  - Class A—Ordinary combustibles
  - Class B—Flammable or combustible liquids
  - Class C—Energized electrical equipment
  - Class D—Combustible metals
  - Class K—Combustible cooking media
- Fire extinguishers that have been tested and approved by an independent laboratory are labeled to clearly designate the classes of fire the unit is capable of extinguishing safely. This traditional lettering system has been used for many years and is still found on many fire extinguishers. More recently, a universal pictograph system, which does not require the user to be familiar with the alphabetic codes for the different classes of fires, has been developed.
- Fire codes and regulations require the installation of fire extinguishers in many areas so that they will be available to fight incipient fires. NFPA 10, *Standard for Portable Fire Extinguishers*, lists the requirements for placing and mounting portable fire extinguishers as well as the appropriate mounting heights.
- Portable fire extinguishers use seven basic types of extinguishing agents:
  - Water
  - Dry-chemicals
  - Carbon dioxide
  - Foam
  - Wet-chemicals
  - Halogenated agents
  - Dry-powder
- Fire extinguishers must always be in a state of readiness. First Responder Inspectors assess the state of readiness by inspecting:
  - The travel distances to the fire extinguisher
  - The mounting of the fire extinguisher
  - Access to the fire extinguisher
  - The type of fire extinguisher

KEY TERMS

Ammonium phosphate An extinguishing agent used in dry-chemical fire extinguishers that can be used on Class A, B, and C fires.

Aqueous film-forming foam (AFFF) A water-based extinguishing agent used on Class B fires that forms a foam layer over the liquid and stops the production of flammable vapors.

Carbon dioxide (CO_2) fire extinguisher A fire extinguisher that uses carbon dioxide gas as the extinguishing agent.

Class A fires Fires involving ordinary combustible materials such as wood, cloth, paper, rubber, and many plastics.

Class B fires Fires involving flammable and combustible liquids, oils, greases, tars, oil-based paints, lacquers, and flammable gases.

Class C fires Fires involving energized electrical equipment where the electrical conductivity of the extinguishing media is of importance.

Class D fires Fires involving combustible metals such as magnesium, titanium, zirconium, sodium, and potassium.

Class K fires Fires involving combustible cooking media such as vegetable oils, animal oils, and fats.

Clean agent A volatile or gaseous fire extinguishing agent that does not leave a residue when it evaporates. Also known as a halogenated agent.

Dry-chemical fire extinguisher An extinguisher that uses a mixture of finely divided solid particles to extinguish fires. The agent is usually sodium bicarbonate-, potassium bicarbonate-, or ammonium phosphate-based, with additives being included to provide resistance to packing and moisture absorption and to promote proper flow characteristics.

Dry-powder extinguishing agent An extinguishing agent used in putting out Class D fires. The common dry-powder extinguishing agents include sodium chloride and graphite-based powders.

Extinguishing agent A material used to stop the combustion process. Extinguishing agents may include liquids, gases, dry-chemical compounds, and dry-powder compounds.

Film-forming fluoroprotein (FFFP) foam A water-based extinguishing agent used on Class B fires that forms a foam layer over the liquid and stops the production of flammable vapors.

Fire load The weight of combustibles in a fire area or on a floor in buildings and structures, including either the contents or the building parts, or both.

Halogenated extinguishing agent A liquefied gas extinguishing agent that puts out fires by chemically interrupting the combustion reaction between the fuel and oxygen.

Incipient The initial stage of a fire.

Loaded-stream fire extinguisher A water-based fire extinguisher that uses an alkali metal salt as a freezing-point depressant.

Multipurpose dry-chemical fire extinguisher A fire extinguisher rated to fight Class A, B, and C fires.

Polar solvent A water-soluble flammable liquid such as alcohol, acetone, ester, and ketone.

Saponification The process of converting the fatty acids in cooking oils or fats to soap or foam.

Underwriters Laboratories, Inc. (UL) The U.S. organization that tests and certifies that fire extinguishers (among many other products) meet established standards. The Canadian equivalent is Underwriters Laboratories of Canada (ULC).

Wet-chemical extinguishing agent An extinguishing agent for Class K fires. It commonly consists of solutions of water and potassium acetate, potassium carbonate, potassium citrate, or any combination thereof.

Wet-chemical fire extinguisher A fire extinguisher for use on Class K fires that contains a wet-chemical extinguishing agent.

You Are the First Responder Inspector

You are inspecting a large industrial occupancy that produces numerous items. The owner recently purchased new extinguishers for the entire property. During your inspection, you encounter numerous types of fire extinguishers, but they are all in incorrect places for the hazards they are meant to protect against. Following your inspection, you present the owner with a list of which fire extinguishers should be used in which locations and for which purpose.

1. The factory area that produces wood shavings and small pieces of lumber:

A. a halogenated-agent extinguisher
B. a Class K extinguisher
C. a 1-B:C-rated extinguisher
D. a pressurized water extinguisher

2. A small, detached, unheated wood storage shed:

A. a 2½-gallon pressurized water extinguisher
B. a 10-B:C-rated extinguisher
C. a Class K extinguisher
D. a 2-A:10-B:C-rated extinguisher

3. The computer server room:

A. Class K extinguishers
B. B:C-rated extinguishers
C. pressurized water extinguishers
D. halogenated-agent extinguishers

4. The kitchen of the cafeteria, which includes deep fat fryers:

A. Class K extinguishers
B. pressurized water extinguishers
C. A:B:C-rated extinguishers
D. B:C-rated extinguishers and Class K extinguishers

CHAPTER

9

Safe Housekeeping Practices

NFPA 1030 THAT INCLUDES CHAPTER 6: FIRST RESPONDER INSPECTOR (NFPA 1031)

First Responder Inspector

- 6.5.10

ADDITIONAL NFPA STANDARDS

- **NFPA 1**, *Fire Code*
- **NFPA 13**, *Standard for the Installation of Sprinkler Systems*
- **NFPA 30**, *Flammable and Combustible Liquids Code*
- **NFPA 31**, *Standard for the Installation of Oil-Burning Equipment*
- **NFPA 33**, *Standard for Spray Application Using Flammable or Combustible Materials*
- **NFPA 54**, *ANSI Z223.1-2015 National Gas Fuel Code*
- **NFPA 80**, *Standard for Fire Doors and Other Opening Protectives*
- **NFPA 101**, *Life Safety Code®*
- **NFPA 115**, *Standard for Laser Fire Protection*
- **NFPA 211**, *Standard for Chimneys, Fireplaces, Vents, and Solid Fuel-Burning Appliances*
- **NFPA 654**, *Standard for the Prevention of Fire and Dust Explosions from the Manufacturing, Processing, and Handling of Combustible Particulate Solids*
- **NFPA 921**, *Guide for Fire and Explosion Investigations*
- **NFPA 1143**, *Standard for Wildland Fire Management*
- **NFPA 1144**, *Standard for Reducing Structure Ignition Hazards from Wildland Fire*
- **NFPA 1405**, *Guide for Land-Based Fire Departments That Respond to Marine Vessel Fires*

KNOWLEDGE OBJECTIVES

After studying this chapter, you will be able to:

1. Describe the importance of good housekeeping practices in fire prevention.
2. List the three requirements of good housekeeping.
3. Identify and evaluate the common fire and life safety hazards related to housekeeping practices outside of buildings:
   - Obstructions to fire protection equipment
   - Fire exposure threats
   - Spontaneous ignition threats
4. Identify and evaluate the common fire and life safety hazards related to housekeeping practices inside of buildings:
   - Spontaneous ignition threats
   - Dust and lint accumulation
   - Mechanical equipment
   - Combustible materials storage
   - Flammable and combustible liquids
   - Painting, coating, and finishing operations
   - Floor cleaning and treatments
   - Fire-protection equipment obstructions
   - Commercial cooking operations
   - Compressed gas cylinders
5. Describe the methods for reducing the risk presented by unsafe housekeeping practices.

SKILLS OBJECTIVES

After studying this chapter, you will be able to:

1. The ability to observe and recognize hazardous conditions.
2. The ability to communicate the conditions observed in an understandable manner.
3. The ability to apply the applicable codes and standards, and make decision in line with the requirements of the AHJ.

You Are the First Responder Inspector

As a new first responder inspector, you are tasked by your department to go out to a factory in your community to conduct a routine inspection. This business is a major employer in town. Upon arrival, you observe a large pile of wooden pallets being stored outside the building, adjacent to numerous doors, windows, and similar building openings. Inside the building you observe that many exit aisles and doors are obstructed or blocked with the storage of raw materials and finished products. The facility's floors have lots of debris—such as cardboard and sawdust—from their process. In addition you see evidence of discarded smoking materials throughout the building.

1. How can you impress upon plant management the need for better control of their housekeeping practices?
2. How much time will it take to correct these items?
3. What strategies should they implement right away and what strategies are worthy of additional time to implement?

Introduction

Good housekeeping practices are an effective fire prevention measure which can best be described as plain common sense. An extensive background in fire safety or fire protection is not needed in order to recognize poor housekeeping practices that potentially pose a fire safety risk. Poor housekeeping practices should serve as a caution to the first responder inspector; places that have poor housekeeping practices often have other fire safety deficiencies (**FIGURE 9-1**).

Safe housekeeping practices—both indoor and outdoor—accomplish four major fire and life safety objectives:

1. Eliminate unwanted fuels, helping to control fire growth and making extinguishment easier.
2. Remove obstructions or impediments to egress.
3. Control sources of ignition.
4. Improve safety for firefighting and emergency response personnel.

Certain aspects of housekeeping are common to almost all types of occupancies while other aspects are unique to certain occupancies. For example, manufacturing occupancies may generate large quantities of combustible dusts that require frequent cleaning. Other occupancies may generate large amounts of trash, waste, or other by-products that need to be removed. In almost all occupancies, the storage of materials can cause obstructions to egress.

As a first responder inspector, you must be able to recognize housekeeping problems and take actions to eliminate them. You can use these opportunities to educate the property owners on the importance of reducing these types of hazards. By taking the time to educate property owners, you will find that the incidence of housekeeping problems will decrease over time.

FIGURE 9-1 Poor housekeeping practices should serve as a caution; places that have poor housekeeping practices often have other fire safety deficiencies.

Housekeeping Overview

A business that keeps its operation neat and clean will have a substantially reduced risk of fire and may benefit in related areas such as lower worker injury rates and improved employee morale. Three basic requirements of good housekeeping are:

- Equipment arrangement and layout—This includes properly cleaning, servicing, and placing devices that generate combustible dusts or waste materials. Sometimes the layout or location of equipment can pose additional problems, such as equipment that is producing dust located close to an ignition source or an air intake.

- Material storage and handling—This includes trash and waste disposal or recycling operations that are found in buildings or on the property outside of a building.
- Operational neatness, cleanliness, and orderliness—This includes emptying trash and waste on frequent intervals to avoid accumulation. The frequency of trash removal may vary substantially depending on the type, form, and amount of trash or waste generated. Some occupancies, such as large retail stores, have very complex and involved trash and waste handling operations to remove, compact, and bale materials for recycling or disposal.

Exterior Issues

You should always inspect the exterior of the property in addition to the inside. In some cases, need for exterior housekeeping may not be apparent to the property owner because these areas are not viewed as often as the inside of the building. Poor housekeeping practices outside of the building can result in obstructions to the site or building, obstructions to fire protection equipment, fire exposure threats, wildfire concerns and an unattractive nuisance easily ignited by vandals or juveniles.

Blocked, Obstructed, or Impaired Access

Exterior housekeeping issues can obstruct access for responding vehicles and firefighters. These access obstructions can delay or hinder firefighting operations. Examples of access obstructions include roads, driveways, fire lanes, or similar accesses being blocked by trash, debris, or dumpsters. Trees or bushes can also become overgrown and block exit doors.

Premise identification, the posting of an address, can aid emergency responders in finding the building. Sometimes the identification numbers may be missing or may not be easily visible from the road. In other cases, the address may have changed. Larger complexes may have separate streets or roads and incorporate their own numbering system. It is critical that fire, law-enforcement, and emergency-medical personnel be able to locate the building quickly in an emergency. For this reason the NFPA 1, *Fire Code* requires that legible address numbers be placed on all buildings. These address numbers should be visible from the road.

PRO TIPS

Different types of occupancies or businesses may have different housekeeping problems depending on the nature and type of processes taking place. Some retail and office occupancies may generate large amounts of combustible trash and recycling materials. Certain industrial and factory occupancies may produce combustible dusts. Other occupancies may have oily rags that need proper disposal.

PRO TIPS

Proper housekeeping procedures are the responsibility of the property owner. As the first responder inspector, you must be able to communicate the need for safe housekeeping practices to the property owner to implement.

In some areas of the United States, snow accumulation can pose additional fire problems. Snow and ice must be removed to make fire hydrants and fire protection equipment accessible. For larger buildings, it may be necessary to plow snow from fire lanes and fire apparatus access roads. One of the most common problems, however, is the failure to keep exits and outside egress paths shoveled or cleared of snow (**FIGURE 9-2**).

FIGURE 9-2 Snow cannot be allowed to accumulate and block an exit.

Obstructions to Fire Protection Equipment

Housekeeping issues can obstruct fire department access to critical firefighting equipment, such as fire hydrants and fire department connections. This can include the presence of trees, bushes, landscaping, or snow that can block fire protection equipment. In some situations, signs can be added to assist firefighters in locating these devices. These signs can identify the location of critical fire protection devices that are otherwise blocked or obstructed from normal view.

In some cases, fire department connections may be obstructed by the addition of a fence (**FIGURE 9-3**). You should ensure that action is taken to ensure that firefighters have access to firefighting equipment by moving or removing the fence or relocating the equipment. In some rare instances where site security is a higher-than-normal concern, you may also consider instructing the property owner to install a gate in the fence with a key located in a secure fire department keybox.

SAFETY TIPS

Obstructed exits also mean obstructed fire department access into and through a building.

Protection of Flammable Liquid and Gas Equipment

When gasoline dispensers, utility gas meters and piping, liquid propane (LP-Gas) tanks, or flammable-liquid tanks are located in areas where they could be struck by vehicles, the fire code requires protection measures to prevent vehicles from impacting this equipment (**FIGURE 9-4**). The lack of impact protection is fairly common with natural gas meters and piping. Protection is commonly accomplished by installing posts or bollards to minimize the possibility of these items being damaged. Other options include installing other barriers—such as guard rails or "jersey barriers"—or installing a high curb to protect these items.

Fire Exposure Threats

Another exterior risk is the accumulation of combustible materials that can pose a fire exposure threat to nearby buildings or equipment. This may involve excessive amounts of trash or debris or may be a by-product of normal business operations. Combustible waste materials from industrial and manufacturing operations are commonly stored on-site before being hauled away. One common example for manufacturing and storage occupancies is the accumulation of idle wood pallets. NFPA 1 has separation requirements for idle pallets (**FIGURE 9-5**). The requirements vary based on the number of pallets and the construction of the building. The minimum separation distance for 50 or more pallets is 30 ft from buildings; this distance is increased to as much as 50 ft if there are relatively large quantities of pallets (over 200). Fencing these outdoor storage areas in order to limit access of unwanted persons is recommended. Ignition sources, such as smoking debris, cutting or welding operations, and hot-work tools such as flame torches, should be kept at a safe distance of 25 ft.

FIGURE 9-3 An obstructed fire department connection is an example of poor outdoor housekeeping.

Courtesy of Jon Nisja.

FIGURE 9-4 If natural-gas meters are not protected against vehicle impact, for example, with bollards, then they and could be easily struck and damaged.

Courtesy of Jon Nisja.

FIGURE 9-5 An excessive pile of wooden pallets could allow fire spread into the building.

Courtesy of Jon Nisja.

FIGURE 9-6 Combustible debris stored next to an LP-Gas tank poses a dangerous hazard for the occupancy.

Courtesy of Jon Nisja.

NFPA 1 requires 35 ft of distance between certain types of hot-work operations and combustible materials. A fire involving these pallets could quickly spread into the facility through the structure's doors and windows.

Another common fire safety problem is the storage of waste-rubber tires outside a building. This is a dangerous practice because this material burns very intensely and can quickly cause severe damage to the building or its contents. NFPA 1 contains limits on the size and number of tires allowed to be stored, in addition to distances that the tires should be kept from buildings.

Combustible waste, rubbish, and debris should never be stored or allowed to accumulate next to flammable or combustible liquid or flammable gas storage tanks or containers. Combustible storage in the proximity of flammable liquid gas tanks or containers can easily be ignited and can cause the failure of the containers and a disastrous fire (**FIGURE 9-6**). Tall grass and weeds should never be allowed to grow near flammable or combustible liquid storage or liquefied propane (LP Gas) installations. A fire involving the grass or weeds would have a disastrous outcome if it reached the tanks containing the combustible product. The use of herbicides should be carefully considered because many of these materials contain chlorate compounds that are oxidizers and can contribute to rapid-fire conditions, especially during dry periods. The use of herbicides is generally limited to pre-emergent grass and weed killers, while any large stand of tall grass or weeds is removed mechanically.

NFPA 1 requires that dumpsters and similar waste receptacles should be located at least 10 ft from combustible buildings and should not be placed under roof eaves or overhangs. If located under eaves or next to combustible buildings or building openings, a dumpster fire can quickly become a building fire. Dumpster fires are relatively common due to discarded ignition materials present in dumpsters, as well as easy access for arsonists.

Tall grass and weeds should never be allowed to grow near flammable or combustible liquid storage or liquefied propane (LP Gas) installations. A fire involving the grass or weeds would have a disastrous outcome if it reached the tanks containing the combustible product. The use of herbicides should be carefully considered because many of these materials contain chlorate compounds that are oxidizers and can contribute to rapid-fire conditions, especially during dry periods. The use of herbicides is generally limited to pre-emergent grass and weed killers while any large stand of tall grass or weeds is removed mechanically.

Wildland Interface

Buildings and facilities constructed near large wildland areas without an intervening firebreak are of growing concern in some areas of the country. These areas are sometimes called a **wildland/urban interface**. Homes, businesses, and apartments have been lost as a result of fast-moving wildland or forest fires. In some jurisdictions, specially developed fire prevention standards or wildland urban interface codes have been adopted to protect against this type of fire spread.

The zone between a structure and an area of native vegetation is known as the wildland-urban interface. Facilities constructed in or adjacent to wildland areas without sufficient **defensible space** or area cleared of combustibles, are at an additional risk of fire. This risk is

FIGURE 9-7 Ladder fuels in the wildland/urban interface area need to be broken.

FIGURE 9-8 Fire department connections blocked by overgrown vegetation are a hazard for firefighters.

a two-directional consideration. First, sufficient defensible space is necessary to stop a wildland fire of trees, brush, grasses, or weeds to spread to outdoor storage or the building itself. Second, a major fire involving a structure or outdoor storage cannot be allowed to spread to the wildland area where a much larger fire may endanger a community.

Breaking the fuel ladder between buildings and outdoor storage and the wildland areas is the essential intent of a defensible space. The **fuel ladder** is a continuous progression of fuels that allows fire to move from brush to limbs to tree crowns or structures. Depending on a number of factors, including **slope**, meaning the upward or downward slant of the land; **aspect**, the compass direction toward which a slope faces; and environmental factors such as weather and winds conditions expected during fire events, the defensible space may need to vary from as little as 30 ft (9 m) to as much as 200 ft (61 m). It is not necessary to remove all vegetation in this space: modifying existing vegetation and fire-safe landscaping treatments appropriate for the climate can achieve a practical solution (**FIGURE 9-7**). Where conditions cannot provide the necessary defensible space, relocating or removing outdoor storage or protecting the structure through improved construction materials, especially roofing, can provide the necessary protection. NFPA 1143, *Standard for Wildland Fire Management*, provides more detailed information on this subject.

Aside from the wildland/urban interface issues, overgrown vegetation, such as weeds, tall grass, brush, shrubs and trees, can cause other concerns for a first responder inspector. These materials can obstruct views or impair access to fire hydrants and fire department connections (**FIGURE 9-8**). Trees and bushes can become so overgrown that they block exit doors and pathways.

Interior Issues

Housekeeping issues inside a building can increase the possibilities of ignition by introducing a heat source, producing larger or more rapidly developing fires due to the additional fuel loads, or can hamper occupant egress or firefighting access by blocking exits and aisles. You should check for general cleanliness in the building. The level of cleanliness is relative to the operation or type of occupancy. Some businesses, such as wood shops, repair garages, and agricultural mills, generate dusts and will be messier than a typical school or office occupancy. A business that keeps its operation neat and clean will have less risk of a fire.

Oily Waste, Towels, or Rags

Spontaneous ignition can occur with oily waste, towels, or rags. **Spontaneous ignition** is the combustion of a material by an internal chemical or biological reaction that has produced sufficient heat to ignite the material. Some of the more common materials subject to spontaneous ignition are linseed oil, Tung oil, charcoal, and certain vegetable oils—most notably peanut oil, which is often used in cooking operations. Oily waste and rags are commonly found in restaurants, for cleaning cooking equipment; vehicle repair garages; industrial occupancies; paint-spraying operations; and building maintenance areas.

Oily waste, towels, and rags should be stored in metal containers with tight-fitting covers. Commercially made containers are available for this purpose. These are metal containers that have a self-closing lid that inhibits fire development by restricting oxygen.

Dust and Lint Accumulation

Accumulations of combustible dust can be a major concern in certain types of manufacturing, storage, and industrial operations. A **combustible dust** is a finely divided solid material that presents a fire or explosion hazard when dispersed and ignited in air. Some dusts pose an extreme explosion risk. Examples of businesses that have a dust explosion risk include agricultural operations where crops are milled or crushed; food product manufacturing involving flour, starches, and sugars; woodworking operations generating sawdust; and coal-handling operations.

It is imperative that these operations minimize dust accumulations by removing dust from equipment and structural components. This should be done by vacuuming or suction; every effort should be made to avoid putting these products in suspension in the air. In most cases, the dust accumulation can be removed using vacuum-cleaning equipment with an explosion-proof motor. Specially made dust collection systems are also manufactured for this purpose. These systems can also incorporate automatic fire detection and suppression systems.

In many cases the dust collection storage vessel, often referred to as the hopper or cyclone, is located outside of the building. You should view the ductwork and hopper to make sure there are no leaks where dust can escape and to minimize exposure to ignition sources. One of the common problems in dust collection systems is the introduction of a spark that causes an explosion. The spark can be from cutting, welding, or grinding operations or from an electrical source.

Processes that generate lint, especially clothes-drying operations, need to be cleaned frequently to minimize build-up of this easily ignitable material. This can especially be a problem with commercial laundry operations that handle large volumes and have dryers heated to higher temperatures to dry clothes more quickly. You should look for evidence of lint accumulation inside the laundry equipment, exhaust ductwork, and in the room itself.

Timber, woodworking, textile or agricultural grain-processing facilities can generate large amounts of combustible dusts or fibers that, when airborne, can explosively ignite or rapidly spread a fire faster than a conventional sprinkler system can control. Dust and fibers can accumulate on walls, ceilings, motors, heating equipment, and structural members; under tables; and inside ducts and conveying equipment. Removing these dusts or fibers can be a dangerous process because an explosive fire can occur if not done correctly.

FIGURE 9-9 A woodworking shop must have a dust collection system to help prevent fires.

In most cases dust, lint, and fibers should be removed by way of a vacuum system employing dust-collection equipment and safe electrical hardware (**FIGURE 9-9**). In a few cases, damp cloths are used to remove dust or fibers, and then are properly disposed in metal containers if spontaneous combustion is a possibility. Compressed air or blowers should never be used to clean dusty areas as this only suspends the material in the air, making it easy to ignite.

Mechanical Equipment

Another interior housekeeping concern is excessive lubrication on motors, engines, compressors, and similar equipment that can attract dirt and dust, resulting in overheating of the equipment. Excessive accumulations of oil, grease, and similar lubrication should be removed from the equipment. If you see oil or grease on motors or compressors, you should require that the property owner remove these accumulations.

Combustible Materials and Storage

Many buildings lack adequate storage provisions, leading people to store materials unsafely. Sometimes, carts or powered equipment used to transport materials within a building can be left in undesirable locations. One of the more common problems seen by first responder inspectors is storage of combustibles and carts in the egress system, causing obstructions and impediments to prompt exiting in an emergency (**FIGURE 9-10**).

Another egress concern is combustible storage under stairways. Should these materials be ignited, the fire can damage the stairs rendering them unusable.

FIGURE 9-10 An exit door that is blocked is a disaster waiting to happen.

In other situations the smoke generated by these burning materials would contaminate the stairs and other egress components, impeding the ability of occupants to use these egress paths. The area under stairs should not be used for storage unless it is separated from the stairs by fire-resistant rated construction.

Storage must also be separated from potential ignition sources, such as boilers, furnaces, water heaters, kilns, heat-producing appliances, and space heaters. In residential settings, such as a nursing home, it is common to find boxes, newspapers, and clothes kept too close to a water heater or furnace. A minimum of 18 in. is specified in NFPA 54, *National Gas Fuel Code*, and NFPA 211, *Standard for Chimneys, Fireplaces, Vents, and Solid Fuel-Burning Appliances*, for gas and electric heaters and 36 in. (914 mm) for high-heat producing appliances, such as boilers, incinerators, and solid-fuel burning appliances.

PRO TIPS

In rack storage configurations, the aisles are typically better defined by the presence of the rack themselves: the racks form the aisles. In solid-piled storage, you may have to work closely with the property owner to define aisles and designated storage areas.

High-Piled Combustible Storage

Many storage and retail occupancies, sometimes referred to as bigbox stores, utilize high-piled storage arrangements to maximize space. **High-piled storage** is often defined as solid-piled, palletized, rack storage, bin box, or shelf storage in excess of 12 ft (366 cm) in height. In some cases the storage arrangement poses risks to firefighting personnel by creating very narrow aisles and unstable piles. This arrangement could hamper firefighting operations, and the piles could collapse on firefighters, especially when boxes became saturated with water or weakened by fire damage.

NFPA 1 contains specific minimum aisle dimensions and maximum pile height and sizes. The minimum aisle dimensions and storage pile dimensions are based on a number of factors, including the commodity being stored and level of fire protection provided. Aisles between racks of storage or solid piles on pallets should be a minimum of 4 ft according to NFPA 13, *Standard for the Installation of Sprinkler Systems*. This distance is needed for safe forklift operations inside the building. In some rack storage conditions, NFPA 13 requires aisles to be a minimum of 8 ft (244 cm).

Another concern that occurs in high-piled storage operations involves safety concerns for products that expand with the absorption of water. Where large quantities of paper products, especially rolled paper, are stored, it is necessary to keep an aisle between the paper storage and the sidewall of the building. During firefighting operations, either from automatic fire sprinklers or firefighting hoselines, these products can soak up water, expand, and push against the outside walls of the building, causing severe structural damage or collapse. For this reason, NFPA 13 requires a 24 in. (610 mm) aisle between exterior walls of the building and materials that will absorb water and expand.

Trash or Recycling Issues

Emptying trash and waste at frequent intervals to prevent accumulation is one example of an effective housekeeping practice. The frequency of trash removal may vary substantially depending on the amount of trash or waste generated. Some occupancies, such as large retail stores or large office buildings, have waste-handling operations to remove, compact, and/or bale materials generated in their everyday operation.

Many states and jurisdictions encourage or require businesses and employers to participate in environmental recycling programs. Recycling programs in many areas encourage or require participation for an increasing number of recyclable materials, such as plastic, and waste considered hazardous to normal waste disposal streams, such as foam products. Although materials involved in recycling programs are not considered trash or waste in the classic sense, they represent the same types of housekeeping issues—they are often combustible and are frequently stored in undesirable arrangements or locations. The number and size of containers for recycling programs can also be much greater. Individual containers for segregation of recycling materials may be located in each work area, with larger collection points in key areas.

Packing and Shipping Materials

Cardboard, paper, styrofoam, expanded plastics, excelsior, straw, and similar materials used for packaging and shipping. Loose packing material, known as **dunnage**, is used to pack, support, and brace products within shipping containers, inside rail cars, on flatbed trucks, and inside a ship's holds. They are considered clean waste and may often be reused or recycled.

All packaging and shipping materials are combustible and represent a significant fire risk. These materials are relatively easy to ignite and have relatively high rates of heat release; some of the plastic materials have rates of heat release similar to flammable liquids. Very large quantities should be protected and kept in separate fire areas.

If the business being inspected uses lots of packaging or shipping materials, these materials should be kept in special fire-rated storage rooms or vaults until they are being used. You should note the condition of shipping and receiving rooms. Pay particular attention to large quantities of accumulated waste near packaging or unloading operations. If this is discovered, a regularly scheduled program for cleaning and waste removal should be instituted immediately. NFPA 1 requires that combustible rubbish not stored in fire-rated rooms or vaults must be removed from the building by the property owner daily. Once removed from the building, these materials should be stored in accordance with requirements for outdoor storage in dumpster or trash areas at a safe distance from the building.

Flammable/Combustible Liquids

Flammable and combustible liquids are commonly found when conducting inspections. The model fire codes typically allow small quantities to be used inside buildings; the amounts will vary with the type of occupancy involved. When not in use, most model fire codes require that flammable and combustible liquids, and other types of hazardous materials, be stored in approved containers or cabinets. A flammable-liquid storage cabinet is used for the storage of flammable and combustible liquids; it protects liquids and their vapors from ignition sources.

Wherever flammable or combustible liquids are handled or used, there is a risk of spills occurring. Businesses that handle flammable or combustible liquids should have an emergency management plan for handling spills and leaks. Both NFPA 1 and federal environmental laws require some sort of hazardous materials management plan. One means of complying with spill control requirements for relatively small quantities of flammable liquids—under 5 gallons (19 L) in maximum anticipated spill quantity—is to have an adequate supply of absorptive materials and tools, which can be used to control, contain, or clean up spills. A common type of absorptive material is a granular product similar to cat litter; it is readily available from automotive parts stores, hardware stores, and safety supply stores.

If larger spills are anticipated, spill control is usually accomplished using some sort of containment system or diking to prevent the flammable or combustible liquids from spreading into the sewer or ground. You should ensure that spill-control measures exist where flammable or combustible liquids are being used.

Disposal of many flammable and combustible liquids has been made more difficult and expensive due to state and federal environmental concerns and laws. Draining into sewers or dumping on the ground are not acceptable disposal methods for combustible materials. Flammable and combustible liquid spills should be cleaned up on site and the materials kept in a separate container for disposal by a hazardous waste disposal contractor. Environmental laws require that they be disposed of by specialized hazardous waste disposal companies.

In certain types of occupancies, such as automotive repair garages and service stations, used motor oil may be stored in tanks on the property. This used motor oil is sometimes referred to as waste oil. It may merely be stored in tanks or drums until it can properly be disposed of or it may be connected to fuel-burning equipment for heating parts of the building. NFPA 1 and NFPA 31 contain specific requirements for waste-oil burners. They can only be installed in industrial occupancies and must be specifically listed to burn waste oil. The tanks must also be listed. Waste oil used in these systems should never contain gasoline because they are designed to burn combustible liquids, such as oil, and not flammable liquids, such as gasoline.

The use of flammable cleaning solvents is becoming fairly rare because of development of nonflammable solvents that have no flash points, are very stable, and have limited toxicity problems. Flammable liquids are still used for some cleaning purposes: examples are alcohols and paint thinners. Flammable liquids used for cleaning should be stored in safety cans with tight-fitting lids that are used only for dispensing small quantities. Flammable liquids should not be stored in open pails, buckets, dip tanks, or containers that may be degraded by the liquid.

As part of a routine fire inspection, you should review or discuss the flammable liquid disposal plans or procedures that the property uses. The plans or procedures often involve contracting with a hazardous waste disposal company.

Paintings, Coatings, Finishes, and Lubricants

Many operations in manufacturing, factory, industrial, vehicle repair garages, and similar businesses use paintings, coatings, or finishes that are sources of combustible residue. A **spray booth** is a power-ventilated enclosure around a spraying operation or process that limits the escape of the material being sprayed, and directs these materials to an exhaust system.

Spray booths, exhaust ducts, fans, and motors need to be cleaned on a regular basis to prevent the dangerous accumulation of these residues. Filters in spray finish operations, such as spray booths and spray rooms, need to be installed and replaced regularly to minimize residue accumulation. These filters trap the residues to prevent them from accumulating in the ductwork and on the fans. Filters with large quantities of residue accumulation that no longer stay in their proper position or that are physically damaged or torn need to be replaced.

Because of the high number of fires associated with spray finishing operations, the area or spray booth is required to have an automatic fire suppression system. You need to note how sprinklers are protected against residue accumulation in spray finish booths and ducts. One method is to cover the sprinkler with cellophane plastic or paper bag that is changed on a regular basis to avoid excessive accumulation or residue.

Floor Cleaning and Treatment

Floor cleaning, treatment, or refinishing can be a fire hazard if flammable solvents or finishes are used. In addition, the removal and refinishing of floor surfaces, especially wood floors, can generate combustible dusts and residues. This risk has been reduced recently because the finishing industry has developed newer products that are not classified as flammable liquids.

If a floor cleaning or refinishing operation is conducted, there should be adequate ventilation, and only

FIGURE 9-11 In sprinkler-protected buildings, NFPA 1 and NFPA 13 require that storage be kept at least 18 in. (457 mm) below sprinklers.

materials having a flash point above the highest room temperature should be used. In addition, nonsparking equipment should be used, and there should be no open flames in the area. The containers for the solvents or cleaning compounds should be labeled as to their flammability or combustibility. If no labels are present, Safety Data Sheets (SDS) should be reviewed.

Some floor treatments and dressings contain oils or compounds that can spontaneously ignite. Oily mops, towels, or rags used to apply these treatments or dressings should be stored in the same manner as oily rags, in metal containers with tight-fitting lids.

Obstructions to Fire Protection Equipment

Improperly stored materials can obstruct access to fire protection equipment, such as fire extinguishers, control valves, and fire alarm pull stations. Improper storage can also impair the proper operation of passive fire protection equipment, such as fire doors.

In sprinkler-protected buildings, NFPA 1 and NFPA 13 require that storage be kept at least 18 in. (457 mm) below sprinklers. This distance allows sprinklers to develop their characteristic "umbrella" spray pattern to effectively extinguish a fire (**FIGURE 9-11**). Storage too close can hamper or impair ceiling-mounted sprinklers.

Kitchen Cooking Hoods, Exhaust Ducts, and Equipment

Most commercial kitchens are equipped with a cooking **hood and exhaust system**. The hood and exhaust

FIGURE 9-12 Excessive accumulations of grease require immediate cleaning.

© vadim kozlovsky/Shutterstock

system are installed above a cooking appliance to direct and capture grease-laden vapors and exhaust gases. Grease accumulation on cooking hoods, on the hoods' grease filters, or inside the exhaust duct represents a serious fire safety risk. This accumulated grease can be ignited from sparks or heat from the cooking operation or from a small fire on the cooking surface.

The kitchen hood and filters should be inspected and cleaned on an established schedule to ensure an excessive amount of grease will not accumulate. Visual inspections will reveal how much grease has accumulated in the ductwork. The amount of grease accumulated will determine how often cleaning will need to occur; it is recommended that the property owner conduct inspections at least weekly. It may have to be inspected and cleaned daily depending on the amount of food cooked and the method of cooking.

The entire hood, grease-removal appliances, exhaust ducts, fans, and related equipment need to undergo a thorough cleaning on a regular basis. This can be a very messy and difficult job, especially in the ductwork itself. Commercial firms that specialize in this type of cleaning should be utilized. You may ask the property owner to produce records showing how often the hood and ductwork is being cleaned. The amount of food cooked and the method of cooking will determine the necessary cleaning frequency. If any grease is observed dripping from the kitchen hood, filters, exhaust duct, or from the exterior of the building, an immediate cleaning is needed (**FIGURE 9-12**). It is also important that filters be in place and clean to ensure proper operation.

Compressed Gas Cylinders

Many industrial facilities, automotive repair garages, and other similar shops need to use **compressed gas cylinders** as part of their operations. One of the most common uses of compressed gas cylinders is an oxygen-acetylene cutting torch for cutting and welding of metals. Compressed gas cylinders are portable pressure vessels of 100 lb (45 kg) water capacity or less, designed to contain a gas or liquid at gauge pressures over 40 psi (276 kPa). There are restrictions on the amounts of certain types of compressed gases, such as flammable gases, permitted in an area. Compressed gas cylinders should also be secured in a manner to prevent them from being knocked over (**FIGURE 9-13**). If the cylinder tips over and the neck of the cylinder ruptures, the cylinder can be propelled like a rocket and do considerable damage.

FIGURE 9-13 A compressed gas cylinder should be secured against being tipped or knocked over.

© francesco survara/Alamy Stock Photo

Control of Smoking

By controlling smoking, a common ignition source can be eliminated. Smoking introduces a flame (lighter or matches) and a smoldering heat source (the cigarette or cigar). Cigarette smoking is generally on the decline in the United States. In addition, many states and communities have enacted smoking regulations that limit smoking in public buildings and workplaces. These factors have led to a decrease in the emphasis that you need to place on the control

FIGURE 14 A cigarette receptacle.

of smoking. It has, however, also led to an increase in people smoking in areas where smoking represents a fire hazard. You should be aware of hidden ash cans or cigarette butts in areas where smoking is prohibited. Outdoor smoking areas should also be surveyed to ensure adequate separation from combustible materials and vegetation, especially in wildland-interface hazard areas.

There are situations where smoking should definitely be prohibited and "No Smoking" signs should be prominently displayed. Smoking should always be forbidden near flammable liquids, both indoors and outdoors; near flammable gases, such as LP Gas and acetylene; and in areas where there are large quantities of combustibles, such as retail and mercantile occupancies. Smoking should also be prohibited in areas where dust accumulations are present, such as woodworking plants, and in areas where there are combustible decorations.

In areas where smoking is allowed, approved smoking receptacles should be provided. Receptacles should not allow the cigarette to fall outside the container and should self-extinguish any combustible materials added to the container. These smoking receptacles can also be filled with sand to assist in extinguishing the smoking materials. Specifically-designed receptacles are available from commercial sources such as hardware stores and safety supply companies (**FIGURE 9-14**).

PRO TIPS

No Smoking Areas

There are definite situations where smoking is prohibited and "No Smoking" signs should be posted prominently. Smoking should always be prohibited indoors or outdoors in close proximity to the following:

1. The storage or use of any flammable or combustible gases, such as acetylene, LP-gas, CNG, or hydrogen
2. The storage, dispensing, or use of any flammable or combustible liquid
3. Locations where organic dust is generated or present, such as grain mills and woodworking shops
4. Locations such as warehouses or stockrooms, where large amounts of combustible materials, including dunnage and other packing materials, are stored
5. Displayed or stored combustible decorations, permanent or seasonal

Public safety and community education programs can also incorporate smoking safety messages for those who choose to smoke. Education on unsafe smoking practices, such as smoking in bed or sneaking a smoke in a hazardous location; support for safer smoking materials; and availability of smoking cessation programs can be communicated to the community as part of an all-risk approach to fire and life safety.

Correction of Housekeeping Issues

Correcting identified housekeeping issues is often a minor expense for the business or property owner. The correction, however, often involves staff time so some owners are sometimes reluctant or slow to make the corrections if it takes staff away from duties that generate income for the business. In the case of repeated, chronic, or on-going housekeeping issues, the long-term solution may involve a change in the internal operation, such as daily removal of all trash, or a remodeling or renovation of the building. The remodeling or renovation may create or add storage space or relocate equipment or operations so that the condition, such as a blocked exit door, no longer exists.

The strategies for correcting housekeeping problems should involve the following considerations:

- Reducing or eliminating sources of ignition:
  - Combustibles too close to ignition sources—A safe separation distance should be maintained between heat sources and combustible materials. The distances are spelled out in NFPA 1. As a rule of thumb, 18 in. (457 mm) of separation is standard from typical heat producing appliances, such as water heaters, furnaces, and stoves. A minimum 36 in. (914 mm) of separation is required from high-heat producing appliances, such as ceiling-mounted heaters and solid fuel-burning appliances.
  - Spontaneous ignition of materials—Materials subject to spontaneous heating should be stored in metal containers with tight-fitting lids. This includes oil-soaked rags, especially those used with tung oil, linseed oil, corn oil, peanut oil, soybean oil, or fish oil. The NFPA's Fire Protection Handbook has an extensive list of materials that are subject to spontaneous ignition.
  - Heat due to friction or overheating—Cleaning and proper maintenance can often prevent these fires from happening. Cleaning away oil, grease, and dirt accumulations allows the equipment to dissipate heat and run at lower temperatures. Lubrication of bearings or moving parts reduces friction and overheating.
- Controlling the fuel load—The less there is to burn, the smaller the fire will be. You should work with the property owner to reduce the amount of waste and combustibles stored in the building. If indoor storage is absolutely necessary, NFPA 1 requires that storage be neat and orderly and that it not obstruct access to exiting. Outdoor storage is usually preferable to inside storage but should still be kept away from building openings and overhangs to prevent an exterior fire from spreading into the building.
- Providing access to the site or building for emergency responders—Premise and building address are critical so emergency responders can find the building. Emergency vehicle access to the site, including parking areas, is critical for firefighting purposes.
- Providing access to fire protection systems and equipment—Firefighters need quick access to fire hydrants, fire sprinkler connections, fire extinguishers, standpipe connections, fire sprinkler controls, and fire alarm control panels. Storage should not be blocking access to these critical fire safety features. First responder inspectors will often find this type of equipment blocked or obstructed. Delays getting to these devices can allow the fire to grow or, in some cases, can cause additional water damage if sprinklers cannot be turned off after a fire has been extinguished.
- Providing and maintaining adequate egress—Life safety should always be your number one priority. Anything that blocks, obstructs, or impedes an occupant's ability to get out of the building quickly in a fire or other emergency needs to be corrected quickly. Storage obstructing egress paths or blocking exit doors are common housekeeping problems.

WRAP-UP

CHAPTER SUMMARY

- Good housekeeping practices are an effective fire-prevention measure that can best be described as plain common sense.
- Three basic requirements of good housekeeping are:
  - Equipment arrangement and layout
  - Material storage and handling
  - Operational neatness, cleanliness, and orderliness
- Poor housekeeping practices outside of the building can result in obstructions to the site or building, obstructions to fire protection equipment, fire exposure threats, wildfire concerns, and an unattractive nuisance easily ignited by arsonists.
- Housekeeping issues inside a building can increase the possibilities of ignition by introducing a heat source, produce larger or more rapidly developing fires due to the additional fuel loads, or hamper occupant egress or firefighting access by blocking exits and aisles. A business that keeps its operation neat and clean will have less risk of a fire.
- The level of cleanliness is relative to the operation or type of occupancy. Some businesses, such as wood

CHAPTER SUMMARY CONTINUED

shops, repair garages, and agricultural mills, generate dust and will be messier than a typical school or office occupancy.

- Spontaneous ignition can occur with oily waste, towels, or rags.
- Accumulations of combustible dust can be a major concern in certain types of manufacturing, storage, and industrial operations.
- Timber, woodworking, textile, or agricultural grain-processing facilities can generate large amounts of combustible dusts or fibers that, when airborne, can explosively ignite or rapidly spread a fire faster than a conventional sprinkler system can control.
- Another interior housekeeping concern is excessive lubrication on motors, engines, compressors, and similar equipment that can attract dirt and dust and result in over-heating of the equipment.
- Storage must also be separated from potential ignition sources, such as boilers, furnaces, water heaters, kilns, heat-producing appliances, and space heaters. In residential settings, such as nursing homes, it is common to find boxes, newspapers, and clothes kept too close to a water heater and furnace.
- The strategies for correcting housekeeping problems should involve the following considerations:
  - Reducing or eliminating sources of ignition
  - Controlling the fuel load
  - Providing access to the site or building for emergency responders
  - Providing access to fire protection systems and equipment
  - Providing and maintaining adequate egress

KEY TERMS

Aspect Compass direction toward which a slope faces. (NFPA 1144)

Combustible dust Any finely divided solid material that is 16.5 miles (420 μm) or smaller in diameter (material passing a U.S. No. 40 Standard Sieve) and presents a fire or explosion hazard when dispersed and ignited in air. (NFPA 654)

Compressed gas cylinders Portable pressure vessels of 100 lb (45.3 kg) water capacity or less designed to contain a gas or liquid at gauge pressures over 40 psi (276 mPa).

Defensible space An area as defined by the AHJ [typically a width of 30 ft or (9.14 m) more] between an improved property and a potential wildland fire where combustible materials and vegetation have been removed or modified to reduce the potential for fire on improved property spreading to wildland fuels or to provide a safe working area for firefighters protecting life and improved property from wildland fire. (NFPA 1144)

Dunnage Loose packing material (usually wood) protecting a ship's cargo from damage or movement during transport. (NFPA 1405)

Fuel ladder A continuous progression of fuels that allows fire to move from brush to limbs to tree crowns or structures.

High-piled storage Solid-piled, palletized, rack storage, bin box, and shelf storage in excess of 12 ft (3.7 m) in height. (NFPA 13)

Hood and exhaust system Devices installed above a cooking appliance to direct and capture grease-laden vapors and exhaust gases.

Premise identification Posting of an address for emergency responders.

Slope Upward or downward incline or slant, usually calculated as a percentage. (NFPA 1144)

Spontaneous ignition Initiation of combustion of a material by an internal chemical or biological reaction that has produced sufficient heat to ignite the material. (NFPA 921)

Spray booth A power-ventilated enclosure for a spray application operation or process that confines and limits the escape of the material being sprayed, including vapors, mists, dusts, and residues that are produced by the spraying operation and conducts or directs these materials to an exhaust system. (NFPA 33)

Wildland/urban interface Any area where wildland fuels threaten to ignite combustible homes and structures. (NFPA 1143)

You Are the First Responder Inspector

You received the list of annual business inspections. On the list is "Barney's Restaurant," "Michelle's Electronics," and "Millennium Sporting Goods."

1. During your inspection of the first business, "Barney's Restaurant," the following item is of significant fire and life safety interest:
   A. The condition and functionality of the door hardware on all egress doors from the dining area and the kitchen.
   B. Evidence of grease accumulations around cooking equipment and kitchen exhaust systems.
   C. Accumulations of empty food-packaging containers and wrappers adjacent to cooking equipment in the kitchen.
   D. All of the above.
2. The least valuable information an inspection of the roof of Barney's Restaurant will provide the inspector is:
   A. A view of surrounding terrain and distant topographic features.
   B. Evidence of any grease accumulations at the discharge of the kitchen exhaust fan.
   C. The general condition of any lightening arrestor system installed on the building.
   D. An indication of the general level of care and preventative maintenance provided to the building by its owner.
3. Upon entering "Michelle's Electronics" you notice a large inventory of older electronic equipment on display. The store has repeatedly marked-down the products hoping for a quick sale. These items are piled up in the aisles and are also partially obstructing the rear exit door. Which of the following concerns you as a first responder inspector:
   A. The business owner may be willing to donate a portion of his inventory to the local school system.
   B. The owner of the business is not concerned with carrying current, state of the art electronics in his inventory.
   C. This is an indication of a struggling business with a large, outdated and expensive inventory and should be watched closely for suspicious fire activities.
   D. The poor housekeeping practices, lending to egress maintenance issues, could lead to injury in the event of an emergency.
4. "Millennium Sporting Goods" has posted a special annual sale in its windows and is opening early to accommodate a large anticipated crowd of shoppers. Which of the following is not a fire and life safety concern:
   A. Prompt removal of shipping cartons and boxes as the special sale goods are unpacked and displayed for sale.
   B. Maintenance of aisles for customers and employees to safely move around the store and access exits.
   C. Enforcement of parking time limits to move traffic in and out of the parking lot.
   D. Checking for the use of damaged electrical to supply special display equipment and fixtures brought in for this sale.

CHAPTER

10

Writing Reports and Keeping Records

NFPA 1030 THAT INCLUDES CHAPTER 6: FIRST RESPONDER INSPECTOR (NFPA 1031)

- 6.3
- 6.3.1
- 6.3.2
- 6.4.2

ADDITIONAL NFPA STANDARDS

- **NFPA 170**, *Standard for Fire Safety and Emergency Symbols*
- **NFPA 13**, *Standard for the Installation of Sprinkler Systems*

KNOWLEDGE OBJECTIVES

After studying this chapter, you will be able to:

1. Describe the common elements of effective written documentation.
2. Discuss the barriers that affect written communication.
3. Describe how field notes, sketches, diagrams, and photographs are used to complete a fire inspection report.
4. Describe the comment elements in all fire inspections reports.
5. Describe the information needed when recording and forwarding complaint findings to the AHJ.
6. Describe how to properly reference a code or standard.
7. Describe how to present evidence during a legal proceeding.

SKILLS OBJECTIVES

After studying this chapter, you will be able to:

1. Prepare internal written correspondence to communicate fire protection and prevention concerns, given a common fire safety issue, so that the correspondence is concise, accurately reflects applicable codes and standards, and is appropriate for the intended audience.
2. Prepare a clear and concise inspection report based on observations from a field inspection.
3. Participate in a legal proceeding, with the findings of a field inspection or a complaint and consultation with the AHJ and legal counsel, so that all information is presented factually and the inspector's demeanor is professional.

You Are the First Responder Inspector

As a new first responder inspector, you are asked to gather information regarding a complaint: a report from a concerned parent about locked and blocked exits at a public school during an after-hours band competition. When you arrive, you find several students milling about the hallways and common spaces, waiting their turn to compete. In an effort to prevent students from roaming the hallways throughout the schools, the maintenance staff has secured various roll-down gates, stairway doors, and exterior exits. The principal tells you that he does not have the staff necessary to patrol every hallway and corridor and that he was told to by the school superintendent to lock certain doors.

1. How will you document your findings and observations?
2. How can you be sure that your report accurately reflects the findings of your inspection?

Introduction

As a First Responder Inspector (FRI), you must be able to document what you observed during the inspection and what needs to be corrected, as well as communicate to the owner or the owner's representative the importance and necessity of correcting any deficiencies found during the inspection. You must also be able to state the code sections that apply and express what the code requires, both verbally and in writing. Finally, you must provide a compliance date based on the severity of the deficiency and the policies of the authority having jurisdiction (AHJ).

It is important that you document the findings and observations of each fire inspection clearly, concisely, and accurately. Every fire inspection should have a written record of the fire inspection. Written records, from standard inspection forms to detailed inspection reports, providing documentation about what existed, what was observed, and what happened, are the basis for corrective actions and verification. Inspection records may also be used in appeals hearings, legal proceedings, training sessions, budget preparation, and code modification meetings.

Since all written communications are official and could be made available to the public, state law requires you to keep them on file for a certain time before they can be discarded. Some items must be kept for three years, until obsolete, or superseded. Others are classified as permanent records and must be kept forever. Since different rules apply to the actual retention periods for different types of records, it is important to know the laws in your state. Fire inspection records

PRO TIPS

Information is valuable only if the meaning is communicated effectively. There are many barriers to effective communication. You can use this simple acronym, **AT WAR**, as a reminder to add value to all communications, both verbal and written.

- **Accurate**—The information must be factual and accurately reflect the situation. Examples include the results of plans review or fire inspection, the applicable referenced codes or standards and the required corrective actions.
- **Timely**—The information must be well-timed. Delays may result in lost opportunities or cost money; hurried reports may be incomplete, inaccurate, or otherwise flawed.
- **Well-presented**—The information must be clear and concise to make sure its meaning is not lost. Use pictures and diagrams where they will help clarify or emphasize a point or observation.
- **Accessible**—All interested stakeholders must be able to have the information. Ask the questions, "Who knows what I need to know?" "What do I know that others need to know?" and "Who else needs to know what I know?" to ensure that no one is left out.
- **Relevant**—The information must be appropriate, in both content and style. It must have meaning for, and be understood by, the intended audience.

should be kept as long as the building exists because fire inspection records are generally subject to open records requests.

Written Documentation

Written records are a vital part of every fire inspection activity. Your written records, from standard inspection forms to detailed inspection reports, are often the only documentation of existing conditions, code violations, and corrective actions that need to be performed. Common uses of written records include:

- Fire inspections, both initial and follow-up
- Appeal hearings
- Legal proceedings
- Other uses, such as fire investigations, staff scheduling and job assignments, training and education, trend analysis, media reports, development of policies and procedures, plan review reports, and budget resource justification

Most written records result from an initial inspection, reinspection, or a complaint. You should prepare a written record for every inspection, even the ones where you are denied entry. Written records document the conditions that existed at the time of your inspection and your observations (**FIGURE 10-1**). It is important to record the following:

- Deficiencies and code violations, and any violations corrected at the time of inspection
- Modifications to systems and equipment
- Maintenance and house-keeping of equipment, processes, or operations
- Hazardous materials storage, handling, and use
- Other hazardous conditions
- Emergency plans, drills, and exercises

Simply stated, a written record provides evidence of any fire and life safety hazards that you identified and the building's degree of code compliance. A written record should clearly and accurately document observed deficiencies and fire hazards, relevant code sections, any required corrective actions, compliance date, and the status of those corrections (in progress and completed).

Barriers That Affect Communication

Written records eliminate one common barrier to communication—trying to remember exactly what was said and by whom. Written records are superior to verbal communications for documentation purposes; however, there are several things that can get in the way of effective communication, even with written records.

Written documentation can be incomplete or inaccurate if observations are not recorded properly, problems overlooked, or the wrong or no code provision is cited. Errors may lead to uncorrected problems or improper corrections. One of the reported factors of The Station nightclub fire in Rhode Island was the lack of documentation about the use of combustible acoustic materials around the stage area. Another was the allegation that the pyrotechnics used had been approved by an authority, but there was no documentation of this in the files.

Any suggestion of personal bias or opinion could affect the perception of the reader and influence how they react. Refrain from inflammatory, offensive, or potentially harassing remarks and from slang (**TABLE 10-1**).

TABLE 10-1

Poor word choice	Preferred word choice
"I told them a thousand times."	"Repeated reminders"
"It was a hole-in-the-wall man cave."	"Small tenant space"
"It was a dumb thing to do."	"Did not meet code requirements"

PRO TIPS

Your written records could be the difference between a successful court case and an unsuccessful one. It is vital that you write clear, accurate, and consistent information on every form of written documentation, from standard inspection forms to detailed fire reports.

Likewise, improper grammar, word usage, or spelling could reflect poorly on you and your agency. If your writing looks unprofessional, it may raise questions about your competency. On the other hand, if your writing is too formal or uses too many "big" words, the reader may not understand the terms used and may tune out your message.

Anything that interferes with the clarity of the message creates a communication barrier. Examples

TOWNSHIP OF LOWER MERION FIRE DEPARTMENT - FIRE INSPECTION REPORT

NOTICE: This inspection is a service provided by the Lower Merion Fire Department. The purpose of this inspection is to ascertain and correct any conditions liable to cause a fire, endanger life from fire, or any violations of the provisions or intent of the Fire Prevention Code of Lower Merion Township.

BLOCK # 12	FIRE DISTRICT # 7	NAME Snyder Pharmacy	EMAIL ron@snyderpharmacy.com
BUSINESS PHONE NUMBER 215-555-1091		ADDRESS 642 N. Main St.	
TYPE OF OCCUPANCY Retail		HEIGHT 14' / SIZE 36 X 100	DWELLING UNITS 0
EMERGENCY CONTACT Ron Snyder		PHONE NUMBER 215-555-1091	STORIES ABOVE GROUND 1 / BELOW GROUND 0
OWNER OR AGENT Ron Snyder		ADDRESS 213 High St,	PHONE NUMBER 215-555-1234

1. ENTRANCES & EXITS	Pass	Fail	N/A	Unk
A. Number of exits	X			
B. Fire escape maintained	X			
C. Exits properly marked	X			
D. Exits properly lighted	X			
E. Exits obstructed or locked	X			
F. Doors open with egress	X			
G. Shaftways marked			X	
H. Fire tower doors closed			X	

2. HOUSEKEEPING	Pass	Fail	N/A	Unk
A. Merchandise stored properly		X		
B. Flammable liquids stored properly	X			
C. Proper rubbish containers	X			
D. Proper aisle space	X			
E. Rubbish in building or exterior	X			
F. High grass or weeds	X			
G. Combustible decorations	X			
H. Oily rags properly stored	X			
I. Chemicals properly stored	X			

3. MAINTENANCE & WIRING	Pass	Fail	N/A	Unk
A. Improperly exposed wiring	X			
B. Overloaded extensions	X			
C. Improper fuses / circuit breakers	X			
D. Broken windows or plaster	X			
E. Storage beneath stairways	X			
F. Combustibles near electrical equip.	X			

4. FIRE EXTINGUISHERS	Pass	Fail	N/A	Unk
A. Inspected	X			
Date 3/currentyr	X			
B. Properly tagged	X			
C. Proper number and type	X			
D. Proper location	X			
E. Properly mounted	X			
F. Hydro-tested	X			
G. Maintained properly	X			
H. Employees trained in use	X			

5. FIRE DOORS	Pass	Fail	N/A	Unk
A. Proper Type	X			
B. Obstructed or locked	X			

6. SPRINKLER SYSTEMS -	Pass	Fail	N/A	Unk
A. Inspected / Date 5/1/currentyr	X			
B. Properly tagged	X			
C. Proper type intakes	X			
D. Intakes marked	X			
E. Intakes capped	X			
F. Intakes blocked	X			
G. Valves accessible & marked	X			
H. Valves open & sealed	X			
I. Heads corroded / dirty	X			
J. Heads obstructed	X			
K. Proper clearance below heads		X		
L. Replacement heads available	X			
M. Wrench available	X			
N. Central station monitored			X	
O. Type (Wet) Dry Combination	X			
P. Hydro static tested / Date			X	
Q. FDC connection:			X	

Location:

7. STANDPIPES - WET DRY	Pass	Fail	N/A	Unk
A. Inspected			X	
Date			X	
B. Properly tagged			X	
C. Proper size intakes			X	
D. Intakes marked and capped			X	
E. Location of connection:			X	

8. HOOD SYSTEMS	Pass	Fail	N/A	Unk
A. Inspected			X	
Date			X	
B. Properly tagged			X	
D. Maintained properly			X	
E. Type K extinguisher			X	
F. Proper location			X	
G. Inspection Log			X	

9. HEATING SYSTEMS	Pass	Fail	N/A	Unk
A. Fuel type nat. gas	X			
B. Properly enclosed	X			
C. Flues and chimney clean	X			
D. Emergency shutoffs available	X			
E. In operating condition and maintained properly	X			
F. Combustibles too close	X			

10. EMERGENCY GENERATOR	Pass	Fail	N/A	Unk
A. Fuel type			X	
B. Flues and chimney clean			X	
C. Inspection / test log available			X	
D. Inspection date			X	

11. FIRE ALARM SYSTEMS	Pass	Fail	N/A	Unk
A. Proper type	X			
B. Operative	X			
C. Proper sounding devices	X			
D. Proper strobe signals	X			
D. Wiring improperly exposed	X			
E. Auto connection				
F. Sending stations properly located and marked			X	
G. Type			X	

12. EMERGENCY LIGHTING	Pass	Fail	N/A	Unk
A. Emergency lights work properly	X			
B. Inspection / test log available	X			

13. PROPERTY STATUS	Pass	Fail	N/A	Unk
A. Fire drills held			X	
B. Log availalable			X	
B. Emergency plan available		X		
C. MSDS sheets available	X			
D. Occupancy signs posted	X			
E. "No Smoking" signs posted	X			
F. Address properly posted	X			
G. Broken windows	X			
H. Sidewalks in good condition	X			
I. Stairs in good condition	X			
J. Handrails	X			
K. Ramps	X			
L. Awnings in good condition	X			

NOTES / CORRECTIONS NEEDED:

Small family owned/operated pharmacy
Rec #1 - Reduce storage to 18" below sprinkler heads
Rec #2 - Post emergency plan (exit path)
Rec #3 - Maintain MSDS in knox box
Revisit in 6 weeks

INSPECTED BY	DATE	RECEIVED BY
[signature]	5/5	[signature]

For LMFD Office Use Only:

FOLLOW UP NEEDED YES NO FIRE CODE VIOLATION ISSUED: YES NO VIOLATION NOTICE NUMBER

FIGURE 10-1 A sample written record.

include the use of long sentences or thoughts that do not lead to clear conclusions. For example, write "The fire extinguishers in Room 146 must be located no more than 75 feet apart ..." instead of, "The fire extinguishers, light blue in color and last serviced by ABC Fire Protection four years ago, are located in all processing rooms, including Room 146, and as described in NFPA 10 (current edition as adopted by the jurisdiction on 1 JAN 2010) have been found to be mounted in places that appear to be inconsistent with distance separations for code requirements for existing buildings or new construction—refer to Reference Photo No. 146-01 as documentation of this deficiency needing correction ..."

The use of technical jargon or acronyms may be confusing to the reader. This is most often a concern in detailed, formal fire inspection reports. Remember who your audience is, and write for your audience. For example, use "expensive" instead of "requires significant capital expenditure" or "fire protection system" instead of "FPS." Avoid the use of jargon and a style that is too structured or formal for the audience. For example, use "lighted for night use" instead of "adequately illuminated for nocturnal operations."

One way to overcome communication barriers is to practice writing the way you speak. This style will come most naturally and if you are an effective speaker, your words will have a ring of truth and sincerity. Be professional, be accurate and concise, but be personable.

Field Notes

Good field notes are the basis for thorough and accurate records. Field notes are the original, on-scene description of conditions that existed at the time of the fire inspection. They document what you observed and serve as reminders to perform additional research and documentation. Good field notes are complete, concise, neat, well written, and free from grammatical and spelling errors. Well-written field notes can make the difference in legal actions. For example, a set of good field notes can be used to re-trace or recreate your actions, support your testimony, and act as evidence.

Most field notes are handwritten, but the growing use of handheld computers provides another way to record field notes and attach digital photographs, maps, and other records (**FIGURE 10-2**). Some first responder inspectors use portable recorders to dictate and voice record their observations. If you voice record your field notes, speak clearly and slowly, be aware of background noises, and keep your comments free from technical jargon, inappropriate or offensive comments, and discussions that do not support the facts (**FIGURE 10-3**).

FIGURE 10-2 Digital cameras can document conditions in the field.

FIGURE 10-3 Some first responder inspectors use portable recorders to dictate and voice record their observations.

Sketches, Diagrams, and Photographs

Sketches, diagrams, and photographs provide visual documentation of observed conditions. Use field sketches as rough drafts to provide graphic representations of room layout and dimensions, egress locations, identified fire hazards, fire protection system features, and similar information (**FIGURE 10-4**). Do not worry about drawing your sketch to scale, but try to keep things in proportion, and write dimensions on the sketch where appropriate. Be sure to record the location of all hazards and noted deficiencies.

Diagrams are the final inspection drawings. Whenever possible, diagrams should be drawn to scale, or include dimensioning measurements and information, and use standard mapping symbols such as those

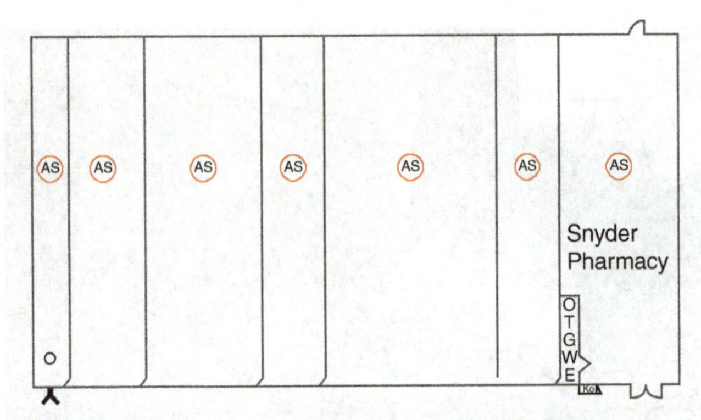

FIGURE 10-4 Use field sketches as rough drafts to provide graphic representations of room layout and dimensions, egress locations, identified fire hazards, fire protection system features, and similar information.

found in NFPA 170, *Standard for Fire Safety and Emergency Symbols*. Be sure to include a legend that illustrates each symbol and its use. Number all drawings so they can be easily referenced and accessed in a written record.

Photographs are effective reminders of what you actually saw during your inspection, and can often convey unsafe conditions better than words can. Photographs are especially helpful when you need proof that a specific condition existed at the time of the inspection. Photographs are often used as evidence in legal proceedings, or to document before and after conditions. Use a date/time stamp for digital photographs to document the time and date of the picture. As with diagrams, number all photographs so they can be easily referenced and accessed.

PRO TIPS

An effective photograph tells a story, accurately and representatively. Two elements of a good photo are composition and lighting. Chapter 15 of NFPA 921, *Guide for Fire and Explosion Investigations*, contains valuable guidelines about how to use photographs to document your observations, and how to improve the photographs you take.

Documenting the Fire Inspection

Fire Inspection Report

The fire inspection report is the most common of all written documentation. The fire inspection report can take several forms, from checklists to detailed reports. Regardless of the format of the fire inspection report, all reports should contain these common elements:

- Be professional and well-written.
- Be focused and concise
- Present facts, free from bias, opinion, or criticism.
- Use correct grammar, word usage, spelling, punctuation, and style. Computer-generated reports and templates can ensure consistent appearance, format, and quality.
- Whenever possible, avoid passive sentences. Use "The property owner shall correct the following" instead of, "The following shall be corrected."

All types of fire inspection reports should contain specific information. The order of the information is not important, but it should be consistent from report to report.

Property Information

Property information describes the building itself. This type of information does not change often. Property information may be used to document and changes to the occupancy size and use. Property information includes:

- Location
- Occupancy type and primary use
- Construction type
- Significant fire protection features, such as fire detection and suppression systems

Contact Information

Contact information is especially useful in multi-tenant occupancies, or when the property owner is located in another city or area of the country and is represented locally by a management company, agent, or other representative. You can use this information in an emergency, to schedule appointments, or to track changes in ownership. Contact information includes the following information for each property owner and tenant:

- Company name and address
- Primary point of contact information, including telephone number, fax number, website, and e-mail address

Inspection-Specific Information

Each fire inspection report should completely and accurately document the findings of the fire inspection. This information includes your name, the name of the property representative, along with the date and time of your inspection. The report should describe the purpose of the inspection and all deficiencies found, including applicable code references. You should record any corrective actions taken or required, and the time allowed for correction to be made. Finally, include any recommendations not required by code at the end of the report and clearly identify them as recommendations to the building owner.

Follow-Up Information

The final part of the report should include follow-up information. Typical follow-up information for the building owner includes reinspection dates, means of requesting additional information or assistance, and a reminder about the importance of completing required corrections. Some state regulations require a notice of right to appeal to be included with any inspection report or notice of violation. Many reports include a closing thank-you statement with the follow-up information.

Reinspection Reports

During reinspections, you do not need to inspect the entire occupancy again. Likewise, you do not need to write a completely new fire inspection report. You should inspect and document only the problem areas noted during your initial fire inspection. Be sure to praise corrective actions and note any remaining deficiencies in your reinspection report.

Additional reinspections may be required, usually when many deficiencies need correcting, when corrections are complex, or when there are multiple correction timelines. If additional reinspections are required, include the planned reinspection date and your contact information.

Always end your reinspection report on a positive note. Thank the property owner or representative for completing any corrective actions and provide a reminder to complete remaining corrections.

PRO TIPS

Remember, reinspection is not a completely new inspection; however, if you identify additional problems, include them in the reinspection report as items separate from the original issues.

Final Notice

A **final notice** should usually be reserved for times that the owner/occupant makes little or no effort to correct problems and deficiencies; however, you should consider using a final notice immediately in exigent, dangerous, or potentially life-threatening situations. Examples of circumstances where a final notice is appropriate include locked or blocked exits, exposed electrical connections, and structural damage caused by fire. The final notice should include consequences of failure to comply with requirements, including legal actions (**FIGURE 10-5**). The first responder inspector will typically work with the AHJ or first responder inspector before a final notice is issued.

Choosing the Right Type of Documentation

It is important to choose the right type of written documentation. You should use the type that is appropriate to the situation, for example, a courtesy reminder of an upcoming annual inspection for a licensed facility, a detailed report of a complex inspection with many observed deficiencies, or a final notice. Some types, such as a courtesy letter, should be friendly and written as if they were spoken. Other types, such as a final notice, require a more formal approach and may require a specific format or presentation to be used.

Letters

Letters are commonly used for final notices and reminders, simple inspection reports, or cover letters for more detailed inspection reports. Each letter you write is a reflection of your agency, so be professional and courteous. It is also important that you write as if you were speaking directly with the person. Develop a personal style that reflects the way you speak, with a focus on the reader. This style will keep your letter personable, yet professional.

Common elements of a successful letter start with a clean, consistent format. Use standard paper size and fonts. A-size, 8.5 in. by 11 in. paper is preferred. Twelve-point type is preferred for legibility; do not use a font size less than 11 point for letters. Every letter should start with salutations and greetings and end with a closing and signature. Make generous use of white space, and start a new paragraph for every subject. The letter should be one page if possible, but should not be longer than two pages. If more space is needed, consider using a checklist or detailed report attached to a cover letter.

A letter and its appearance are a reflection of your agency. Use a consistent format for all agency letters. Spell out the month in every date. Use formal salutations such as Mr., Ms., and Dr. where appropriate. Use a

Township of Upper Moreland

Office of tlhe Fire Marshal

117 Park Avenue • Willow Grove, Montgomery County, Pennsylvania 19090
Phone: 215-659-3100 • Fax: 215-659-1364 • Email: uppermoreland.org

FIRE CODE INSPECTION REPORT

OCCUPANCY:		CONTROL:		
ADDRESS:		OWNER/MGR/PRINCIPAL:		
		ADDRESS:		
PHONE:				
ICC USE:	INSPECTOR:	DATE:	DAY:	TIME:

✔VIOLATION SEE REMARKS ↓

❑ PERMIT INSPECTION
❑ BLOCK INSPECTION
❑ RESIDENTIAL INSPECTION
❑ TANK INSTALL/REMOVAL
❑ FOLLOW-UP

✔VIOLATION SEE REMARKS ↓

❑ TEST ____
❑ COMPLAINT ____
❑ OCCUPANCY ____
❑ MAR ____
❑ OTHER ____

1. Fire Extinguishers due to be inspected.
2. Fire Extinguishers not provided/installed properly.
3. Keep EXIT signs illuminated and visible.
4. Provide additional EXIT signs (high and low).
5. Maintain adequate aisle width.
6. Keep fire exits unlocked and free of obstructions.
7. Repair/install panic hardware.
8. Removed obstructions from fire towers or escapes.
9. Maintain egres slighting.
10. Provide/maintain emergency lighting.
11. Maintain 24″clearance at ceiling with combustibles.
12. Repair voids in the ceiling: holes, tiles, etc.
13. Remove combustible storage from boiler room.
14. Maintain proper housekeeping.
15. Remove grease from ranges, hoods, duct, fans, etc.
16. Keep electrical appliances clear of combustibles.
17. Repair improper wiring, fuses, grounding, etc.
18. Provide proper storage of compressed gas cylinders.
19. Oxidizers stored separately from flammable liquids, corrosive liquids, combustible materials.
20. Provide proper containers/storage of flammable liquids.
21. Bond wires shall be used when dispensing Class I or II liquids from metal to metal containers.
22. Discontinue smoking and post proper signs.
23. NFPA signs posted at entrance to buildings where hazardous materials are used or stored.
24. Provide/maintain fire zone signs
25. Install temporary fire zone signs.
26. Discontinue Improper/Illegal burning.
27. Dumpster/combustibles too close to building.
28. Post address on building/sign.
29. Street address posted on rear of strip occupancy.
30. Occupancy limit sign(s) missing/improperly posted.
31. Failure to obtain required permit.
32. Required test or drill logs missing/incomplete.
33. Required emergency plan missing oroutofdate.
34. Fire alarm system annual test report.
35. Fire sprinkler system annual test report.
36. Sprinklers obstructed.
37. Other suppression system test report. (List in remarks.)
38. Knox Box Key Check.
39. Other (List in remarks).

REMARKS: ____

❑ Haz Mati Non-Permit/Req'd. Site ❑ Sprinkler ❑ Standpipe ❑ Cooking System ❑ AFA ❑ UST/AGT

IN THE INTEREST OF FIRE SAFETY AND TO COMPLY WITH THE UPPER MORELAND TOWNSHIP FIRE CODE, THE ABOVE VIOLATIONS MUST BE CORRECTED IMMEDIATELY. FAILURE TO COMPLY WILL RESULT IN PENALTIES AS SET FORTH IN THE FIRE CODE OF UPPER MORELAND TOWNSHIP. ARE INSPECTION WILL BE ON OR ABOUT 30 DAYS. RECEIPT OF NOTICE ACKNOWLEDGED:

SIGNATURE: ____ INSPECTOR: ____ BADGE: ____
PRINT NAME: ____ FIRE MARSHAL: ____ DATE: ____

THIS REPORT IS BASED UPON OBSERVATIONS AT THE TIME OF SURVEY WHICH MAY NOT DISCOVER ALL HAZARDS.

FIGURE 10-5 The final notice should include consequences of failure to comply with requirements, including legal actions.

subject or reference line to help your reader understand what the letter is about. Each letter should include these basic design elements:

Header	Agency letterhead or return address Date Inside address Subject/reference
Salutation	Dear Mr., Ms., or other recognized and appropriate form
Body	Start with an introduction, followed by the main topic(s) of the letter and ending with a closing paragraph or statement. Start a new paragraph for every major point.
Closing/Signature	Begin with "Sincerely" or "Regards," and end with your signature, printed name, and title.
Enclosures	List all documents or other information included with your letter.
Copies	List the name of everyone who received a copy of this letter, followed by their company or agency name.

Email

A documentation tool that is growing in popularity is email. Most often, emails are used to respond to a request for information or code interpretation. Email can be a convenient, useful, and inexpensive way to communicate with property owners, design professionals, and others. Not quite as immediate as a telephone call, but more timely and less expensive than a detailed report delivered by post or courier, emails are commonly used by FRI; however, you should remember that emails are considered legal documents, work products, or official records subject to the Freedom of Information Act and similar state Public Information Act laws. Your email correspondence should be professional and follow the same standards as your correspondence via letters. During legal proceedings, emails can be subpoenaed. Thus all emails that contain information that is relevant to your work should be saved according to your agency's records retention policies, just like any other written document.

Checklists

Common uses for checklists include simple fire inspection forms that list common and/or critical deficiencies and preliminary inspections of multiple locations with similar characteristics, such as schools or roadside vendors. One advantage of checklists is that they can be used during a fire inspection as a reminder of common issues. However, it is easy to forget or overlook less common problems, so you must be careful that you inspect "to code" rather than "to the checklist."

You can use a checklist as part of your post-inspection interview, and leave a copy of the checklist with the owner (**FIGURE 10-6**). In many cases, a well-designed checklist is the only inspection record needed. In some cases, you may want to include diagrams or photographs to illustrate specific points of your inspection. You can also follow-up the checklist with a formal written report if needed.

A successful checklist starts with a clean, consistent format with easy-to-read groupings. Use standard paper size and fonts. A-size, 8.5 in. by 11 in. paper is preferred. Twelve-point type is preferred for legibility; but smaller fonts, down to 8 point, are acceptable. Be sure to check the font type and size for legibility. Margins and white space are not as critical for checklists. Margins often are as close to the edge of the page as a printer or photocopier will allow, usually 0.5 in. However, remember to leave adequate margin for binding or hole punching to ensure that critical information is not lost.

Checklists are commercially available, but you may want to develop your own checklists. Developing your own checklists and forms allows you to customize them to meet your needs based on the types of occupancies and conditions that are common in your community.

Detailed Reports

Detailed reports should be used to document serious deficiencies, information that will not fit easily into a letter or checklist format, numerous violations, or corrective actions required. You should also consider a detailed report when multiple occupancies or multiple tenants are part of the same inspection process. Detailed reports often include field notes, sketches, diagrams, and photographs.

Reports may be handwritten. A handwritten report may appear less threatening or informal, and lends an air of immediateness to an inspection. As an added advantage, you can leave a copy with the owner after the inspection; however, a handwritten report may appear poorly written, sloppy, or ignorable.

A typed report can be more time consuming to prepare but appears more official and authoritative. Most formal reports are accompanied by a cover letter and support documentation, such as diagrams, photographs, equations, and detailed analyses.

An effective report should use a clean, consistent format with easy-to-read groupings. Use standard

Inspection Checklist
Inspection Procedures

PREINSPECTION CHECKLIST

Equipment: ______

General
- ❑ Identification (photo ID)
- ❑ Business work hours

Clothing
- ❑ Coveralls
- ❑ Overshoes
- ❑ Boots

Personal Protective Equipment (PPE)
- ❑ Hard hat
- ❑ Safety shoes
- ❑ Safety glasses
- ❑ Gloves
- ❑ Ear protection
- ❑ Respiratory protection

Tools
- ❑ Flashlight
- ❑ Tape measure(s)
- ❑ Pad (graph paper) and pen or pencil
- ❑ Magnifying glass

Test gauges
- ❑ Combustible gas detector
- ❑ Pressure gauges
- ❑ Pitot tube or flow meter

Plans and Reports
- ❑ Previous reports
- ❑ Violation notices
- ❑ Previous surveys
- ❑ Applicable codes and standards

Notes: ______

SITE INSPECTION

Property Name: ______

Address: ______

Occupancy Classification
- ❑ Assembly
- ❑ Educational
- ❑ Day care
- ❑ Health care
- ❑ Ambulatory health care
- ❑ Detention and correctional
- ❑ One- and two-family dwelling
- ❑ Lodging and rooming
- ❑ Hotel/Motel/Dormitory
- ❑ Apartment
- ❑ Residential board and care
- ❑ Mercantile
- ❑ Business
- ❑ Industrial
- ❑ Storage
- ❑ Mixed

 (Page 1 of 2)

FIGURE 10-6 A sample checklist.

Hazard of Contents

- ❑ Light (low)
- ❑ Ordinary (moderate)
- ❑ Extra (high)
- ❑ Mixed
- ❑ Special hazards

Exterior Survey

- ❑ Housekeeping and maintenance

Building construction type

- ❑ Type I (fire resistive)
- ❑ Type II (noncombustible)
- ❑ Type III (ordinary)
- ❑ Type IV (heavy timber)
- ❑ Type V (wood frame)
- ❑ Mixed

Construction problems

Building height ____________ feet ____________ stories

- ❑ Potential exposures
- ❑ Outdoor storage
- ❑ Hydrants

Fire department connection

- ❑ Vehicle access
- ❑ Is it obstructed?
- ❑ Is it identified?
- ❑ Drainage (flammable liquid and contaminated runoff)
- ❑ Fire lanes marked

Building Facilities

- ❑ HVAC systems
- ❑ Electrical systems
- ❑ Gas distribution systems
- ❑ Refuse handling systems
- ❑ Conveyor systems
- ❑ Elevators

Fire Detection and Alarm Systems

See Form A-8.

Fire Suppresion Systems

See Form A-10.

Closing Interview

- ❑ Imminent fire safety hazards
- ❑ Maintenance issues
- ❑ Housekeeping issues
- ❑ Overall evaluation

Items to be researched:

- ❑ ____
- ❑ ____
- ❑ ____

Report

- ❑ Draft
- ❑ Review
- ❑ Final

Notes: ____

____

____

____

(Page 2 of 2)

FIGURE 10-6 *(Continued)*

paper size and fonts. A-size 8.5 in. by 11 in. paper is preferred. Twelve-point type is preferred for legibility; but smaller fonts, down to 8 point, may be appropriate is some cases. Check the font type and size for legibility. Long or detailed reports often use tabs and appendices to improve readability, provide additional reference information, or to provide a common place for tables, figures, charts, and photographs.

Handling Assigned Complaints

A complaint occurs when a member of the public indicates that there is, in their opinion, a safety issue at a building. Complaints are typically assigned to the first responder inspector. Your role in this capacity is to gather facts by investigation and then forward them to the AHJ, as the AHJ's policy dictates. If you determine that there are no code issues during your investigation, then the complaint is unfounded. If you note an issue that must be corrected, you will need to document the issue properly.

PRO TIPS

Complaint inspections do not take the place of any scheduled inspections. The complaint requires only looking for one specific item, but other violations that can be seen should also be included in any complaint investigation. The routine inspection is a full top-to-bottom look at the building.

Regardless of whether a specific complaint form is used or not, certain information should be recorded. You should date when your complaint inspection is conducted and document the results of your inspection. You should also document the conditions found, accurately and factually. As mentioned earlier in this lesson, photographs can be extremely helpful for the AHJ to render a decision on what actions to take next.

When conducting the reinspection, the date of the reinspection should be noted, as well as what was found. If the violation is corrected, the complaint form is now completed, and it can be closed.

Code References

The proper way to document a violation is to indicate both the violation and what must be done to correct it. While it may be correct to state that the rear door does not meet code requirements, it is more helpful to indicate that deadbolts must be removed or the door cannot be blocked. To give the owner an opportunity to thoroughly investigate the violation, a code reference should be given. A typical reference might look like "This violates NFPA 13 section 10.3.2" In the cases when violations are noted and fixed immediately during the inspection, the code reference would not be needed; however, the violation should be documented, and you should indicate that the issue was resolved during inspection.

PRO TIPS

In the codes, you will notice the use of the following words, "shall, and is authorized to." "Shall" is mandatory, and is used when an action is required; no latitude or discretion is allowed. Likewise, use of the term "is authorized to" indicates that you have the power to do something but are not obligated to do so.

Record-Keeping Practices

Every fire inspection should include written documentation about the inspection. The documentation should, at minimum, note existing conditions and the correction of, or failure to correct, identified violations. Each piece of documentation should become part of the permanent record about a particular building or occupancy (**FIGURE 10-7**). The inspection files are an important tool to improve fire safety and, if needed, can be used in legal proceedings.

FIGURE 10-7 Each piece of documentation should become part of the permanent record about a particular building or occupancy.

PRO TIPS

All inspection files, with few exceptions, are official government documents. This includes forms, checklists, detailed inspection reports, field notes, e-mails, sketches, photographs, and related materials.

PRO TIPS

"Write each report like it will end up in court, and that it will be published on the front page of the newspaper, because it might."—Chief Raymond Parent

PRO TIPS

Do not take legal proceedings personally.

Participating in Legal Proceedings

The first responder inspector is often called on to participate in legal proceedings, based upon inspections they have performed. A code violation begins with a notation on a fire inspection form and may end in a courtroom. If you have poor documentation, presenting concrete material in court may be difficult; however, with good, detailed documentation, testimony will be supported by what you wrote and photographed during the fire inspection.

Whenever you conduct a fire inspection, all documentation and correspondence is subject to use in a legal proceeding, so document thoroughly. You never know when findings from a routine inspection may end up in court. If a fire or injury occurs in the building, all of its records, including past inspection reports, will be pulled. Everyone from fire officials, building officials, government officials, and attorneys will be examining the records to see what was noted on previous inspections. They may be looking for a trend of recurring violations or a violation that was never documented.

Presenting Evidence

If you have to go to court to present evidence, you should review all a building's records until you are well-versed on that building, from type of occupancy to any violations listed during the its lifetime. Review the codes to make certain that you completely understand exactly what the code requires. For example, say you are testifying about a fire code violation that has not been corrected, even after numerous inspections. Due to poor housekeeping conditions, the fire could spread to adjacent buildings. Before the court case, you would examine the records to see if the owner had a history of violations.

When in court, act and look professional. Wear your uniform or professional attire. The witness stand is not the stage to be clever or witty; simply answer the questions you are asked. For example, if you are asked, "Do you know what day it is?" your answer should be "Yes," not "It's Monday, July 10." If the attorney wants to know the date, the attorney will ask, "What is the date?"

When giving testimony, only state information you know is fact. If you cannot remember a fact, simply state that you are uncertain or that you need to refresh your memory and would like to look at your paperwork. If you do not know the answer to a question, say that you do not know. Keep your answers short. There is no need to expand on a yes or no answer, unless the answer needs clarification. Refrain from giving your opinion—you are in court as a fire official to present the facts.

WRAP-UP

CHAPTER SUMMARY

- As a first responder inspector, you must be able to document what you observed during the inspection and what needs to be corrected, as well as communicate to the owner or the owner's representative the importance and necessity of correcting any deficiencies found during the inspection.
- Your written records, from standard inspection forms to detailed inspection reports, are often the only documentation of existing conditions, code violations, and corrective actions that need to be performed.
- Written records eliminate one common barrier to communication—trying to remember exactly what was said and by whom. Written records are superior to verbal communications for documentation purposes.

CHAPTER SUMMARY CONTINUED

- Good field notes are the basis for thorough and accurate records. Field notes are the original, on-scene description of conditions that existed at the time of the fire inspection.
- Sketches, diagrams, and photographs provide visual documentation of observed conditions. Use field sketches as rough drafts to provide graphic representations of room layout and dimensions, egress locations, identified fire hazards, fire protection system features, and similar information. Diagrams are the final inspection drawings. Photographs are visual evidence of what was seen during the inspection.
- All fire inspection reports should include property information, contact information, inspection-specific information, and follow-up information.
- Reinspection reports often just need to document if the violations found during the initial inspection were properly addressed.
- A complaint occurs when a member of the public indicates that there is, in their opinion, a safety issue at a building.
- To give the owner an opportunity to thoroughly investigate the violation, a code reference should be given.
- A code violation may begin with a notation on a fire inspection form and may end in a courtroom. If you have poor documentation, presenting material in court may be difficult, as you must rely almost completely on your memory.

KEY TERMS

Final notice Written correspondence used when violations are not corrected to notify the owner that legal action may be taken to ensure code compliance.

You Are the First Responder Inspector

It is your first day as a first responder inspector. You have been assigned a department vehicle and issued the personal equipment necessary to perform routine fire inspections. The inventory list of personal equipment includes the following:

- Camera, digital
- Checklists for common occupancy locations
- Clipboard, with ruled and graph paper, pencils, and ruler
- Measuring tape, 100-ft
- Hard hat, leather gloves, latex gloves, and safety glasses
- Portable radio and cellular telephone
- Local codes and standards

1. Later that day, you are assigned to complete an initial fire inspection of an existing daycare center. Which of the following pieces of equipment will you most likely leave in your car during this visit?

A. Checklist listing the common violations in daycare centers
B. Digital camera
C. Clipboard, with ruled and graph paper, pencils, and ruler
D. Local codes

2. While you inspect the daycare center, you notice a burned-out light bulb in an exit sign above an emergency exit door. There are no other violations. You should:
   - **A.** Document the deficiency, the required corrected action, and the follow-up inspection date, all in writing.
   - **B.** Ignore it; it's not necessary to correct if only one problem exists.
   - **C.** Close the daycare and arrest the operator for a serious code violation.
   - **D.** Verbally explain the problem and verbally ask the operator to correct it before the next annual visit.
3. On your annual visit a year later, you discover a new addition with two rooms and a kitchen, but no commercial vent or suppression system. None of these modifications is shown on the original site plans in the permanent records. You should:
   - **A.** Issue a Fire Marshal's Order to supply architect-certified drawings for the additions.
   - **B.** Use field notes, sketches, and photographs to document the additions as part of your inspection process.
   - **C.** Use a simple cover letter to document the inspection and required actions.
   - **D.** Both A and B.
4. You check your e-mails and find a request from the local television station's investigative reporter. He says he has received a call from a parent of a child at the daycare that has complained about "severe fire hazards" at the daycare. He asks for an on-camera interview and for you to send him a copy of "any and all records of any and all inspections." He includes his contact information, including cell phone number. You should:
   - **A.** Ignore the request because the e-mail request is not on station letterhead.
   - **B.** Call the reporter and ask why he wants the information.
   - **C.** Follow your department's open records request policies and procedures.
   - **D.** Agree to the on-camera interview, but do not provide any written reports.

Appendix A

NFPA 1030 Standard for Professional Qualifications for Fire Prevention Program Positions, 2024 Edition that includes Chapter 6: First Responder Inspector (NFPA 1031) Correlation Grid

Chapter 6: First Responder Inspector (NFPA 1031)

6.1 Administration

6.1.1 Scope Chapters 6 through 8 identify the minimum job performance requirements (JPRs) for first responder inspector, fire inspector, and fire plans examiner.

6.1.2 Purpose The purpose of Chapters 6 through 8 are to specify the minimum JPRs for serving as a first responder inspector, fire inspector, and fire plans examiner.

6.1.2.1 Chapters 6 through 8 shall not address management responsibility.

6.1.2.2 It is not the intent of Chapters 6 through 8 to restrict any jurisdiction from exceeding or combining these minimum requirements.

6.1.3 General

6.1.3.1 The first responder inspector, fire inspector, or fire plans examiner candidate shall be skilled in written and oral communications, public relations, and basic mathematics.

6.1.3.2 The first responder inspector, fire inspector, or fire plans examiner candidate shall meet the JPRs of Section 5.2 of NFPA 470.

6.1.3.3 The JPRs for each level of progression shall be completed in accordance with recognized practices and procedures or as defined by law or by the AHJ.

6.1.3.4 The JPRs need not be mastered in the order in which they appear.

6.1.3.5 The local, state/provincial, or federal training programs shall establish the instructional priority and the training program content to prepare individuals to meet the JPRs of Chapters 6 through 8.

6.1.3.6 Evaluation of JPRs shall be by individuals approved by the AHJ.

6.1.3.7 Prior to appointment or assignment to a position, personnel shall meet the following requirements as established by the AHJ:

(1)	Educational requirements
(2)	Age requirements
(3)	Medical requirements
(4)	Job-related physical performance requirements
(5)	Background investigation requirements
(6)	Character traits as required by the position

6.1.3.8 A person assigned the duties of first responder inspector shall meet all of the requirements defined in Chapter 6 prior to being qualified as a first responder inspector.

6.1.3.9 A person assigned the duties of fire inspector shall meet all of the requirements defined in Chapter 7 prior to being qualified as a fire inspector.

6.1.3.10 A person assigned the duties of fire plans examiner shall meet all of the requirements defined in Chapter 8 prior to being qualified as a fire plans examiner.

6.1.3.11 The first responder inspector, fire inspector, and fire plans examiner, who performs or supports the duties and responsibilities covered by this standard shall remain current with the required requisite knowledge, requisite skills, and individual JPRs addressed for each level or position of qualification to maintain proficiency and competency with the JPRs covered in this standard.

6.1.3.12 The first responder inspector, fire inspector, or fire plans examiner shall perform assigned duties in accordance with applicable safety standards.

6.1.3.13 The AHJ shall provide personal protective clothing and the equipment necessary to conduct assigned inspections and plan review.

6.1.3.14 The first responder inspector, fire inspector, or fire plans examiner shall be provided with codes and standards, and the policies and procedures applicable to the AHJ and their assignment.

6.1.3.15 The first responder inspector, fire inspector, and fire plans examiner shall complete inspections, plan review duties, and perform other related activities, so that available time is used efficiently.

6.1.3.16 The first responder inspector, fire inspector, and fire plans examiner shall be able to develop written correspondence to communicate fire protection and fire and life safety code requirements, so that the correspondence provides an accurate interpretation of applicable codes and standards and is for the intended audience.

6.1.3.17 The first responder inspector, fire inspector, and fire plans examiner shall maintain records and related documents, so that information can be retrieved and filed in compliance with the record-keeping policies of the AHJ.

6.1.3.18 The first responder inspector, fire inspector, and fire plans examiner shall be able to read plans in a format acceptable to the AHJ.

Chapter 6: First Responder Inspector (NFPA 1031) JPRs	Chapter(s)	Page(s)
6.2 General The first responder inspector shall meet the JPRs defined in Sections 6.2 through 6.6. **6.3 Administrative Duties** This duty involves the preparation of inspection reports, handling of complaints, and maintenance of records, as well as maintaining dialogue with fire inspectors and fire plans examiners and other relevant personnel, according to the JPRs in Sections 6.3.1 through 6.3.6. **6.3.1** Prepare internal written correspondence to communicate fire protection and prevention concerns, given a common fire safety issue, so that the correspondence is concise, accurately reflects applicable codes and standards, and is appropriate for the intended audience.		
6.3.1(A) Requisite Knowledge Applicable policies of the AHJ. **6.3.1(B) Requisite Skills** Communication methods as prescribed by the AHJ.	1, 10	4–9, 164–175
6.3.2 Prepare inspection reports, given AHJ policy and procedures, and observations from a field inspection, so that the report is clear and concise and reflects the findings of the inspection in accordance with the applicable codes and standards and provides actions required based on the policies of the AHJ.		
6.3.2(A) Requisite Knowledge Applicable policies, codes, and standards adopted by the AHJ **6.3.2(B) Requisite Skills** The ability to conduct a field inspection, apply AHJ policy, and communicate orally and in writing.	10	164–175
6.3.3 Identify the applicable code or standard, given fire protection, fire prevention, or life safety deficiencies observed during an assigned fire inspection, so that the applicable document, edition, and section are referenced.		
6.3.3(A) Requisite Knowledge Fire and life safety codes and standards, and policies as determined by the AHJ. **6.3.3(B) Requisite Skills** The ability to apply codes, standards, and policies as determined by the AHJ.	1, 4, 6, 7, 8, 10	4, 52, 59, 63–64, 99, 123, 143, 174
6.3.4 Recognize the need for a permit, given a situation or condition, so that requirements for permits are communicated to the building owner, owner's representative, occupant, event organizer, and fire prevention staff in accordance with the policies of the AHJ.		
6.3.4(A) Requisite Knowledge Permit policies of the AHJ and the rationale for the permit. **6.3.4(B) Requisite Skills** The ability to communicate orally and in writing.	1	8
6.3.5 Investigate assigned complaints, given a reported situation or condition, so that complaint information is recorded, and the findings are forwarded to the AHJ in accordance with AHJ policy.		
6.3.5(A) Requisite Knowledge Applicable policies of the AHJ. **6.3.5(B) Requisite Skills** The ability to recognize problems, apply fire prevention principles, communicate orally and in writing, and forward as required to the AHJ.	4	50, 53, 56, 64
6.3.6 Identify fire and life safety hazards or conditions, given a fire protection, fire prevention, or life safety issue, so that the applicable action is taken per AHJ policy.		
6.3.6(A) Requisite Knowledge Fire and life safety hazards and the applicable codes and standards and the policies of the AHJ. **6.3.6(B) Requisite Skills** The ability to apply codes, standards, and policies.	4	58–62

Chapter 6: First Responder Inspector (NFPA 1031) JPRs	Chapter(s)	Page(s)
6.4 Legal **6.4.1** This duty involves the knowledge of various legal proceedings such as enforcement of the adopted codes and standards of the AHJ, handling various complaints, and initiating legal action where necessary. **6.4.2** Ability to participate in legal proceedings, given the findings of a field inspection or a complaint and consultation with the AHJ and legal counsel, so that all information is presented factually and the inspector's demeanor is professional.		
6.4.2(A) Requisite Knowledge The legal requirements pertaining to evidence rules in the legal system and types of legal proceedings in accordance with the AHJ. **6.4.2(B) Requisite Skills** The ability to maintain a professional demeanor, communicate, listen, and differentiate facts from opinions.	1, 4, 10	5, 6, 56, 64, 174, 175
6.5 Field Inspection This duty involves conducting assigned fire safety inspections of existing structures and properties to identify fire and life safety hazards, according to the JPRs in Sections 6.5.1 through 6.5.10. **6.5.1** Determine code compliance, given the codes and standards, the policies of the AHJ, and a fire protection issue, so that the applicable codes, standards, and policies are identified and compliance is determined.		
6.5.1(A) Requisite Knowledge An understanding of inspection practices and applying code requirements. **6.5.1(B) Requisite Skills** The ability to observe, recognize, and report problems.	1, 4, 10	4, 52, 58–64, 99, 123, 143, 174
6.5.2 Identify the fire and life safety hazards, given an existing occupancy, so that violations are identified based on a specific occupancy.		
6.5.2(A) Requisite Knowledge Fire and life safety hazards by occupancy. **6.5.2(B) Requisite Skills** The ability to make observations, identify violations, and forward observations and hazards to the AHJ.	2, 4	12–23, 51–61
6.5.3 Verify occupancy classification of a single-use occupancy, given a description of the occupancy and its use, so that the classification is made according to the applicable codes and standards.		
6.5.3(A) Requisite Knowledge Occupancy classification types adopted by the AHJ. **6.5.3(B) Requisite Skills** The ability to observe, recognize, and report problems.	2, 4	12–23, 51
6.5.4 Verify that the means of egress elements are maintained, given an existing occupancy, so that the elements are free of obstructions; easily operated; not locked; and that deficiencies are identified, documented, and reported in accordance with the applicable policies of the AHJ.		
6.5.4(A) Requisite Knowledge Applicable knowledge related to means of egress elements, and maintenance requirements of egress elements. **6.5.4(B) Requisite Skills** The ability to observe and recognize problems, and make decisions related to means of egress.	5	69–80
6.5.5 Verify posted occupant load, given an occupancy classification, so that a building or structure is occupied in accordance with applicable codes and standards and policies of the AHJ.		
6.5.5(A) Requisite Knowledge An understanding of occupant loads. **6.5.5(B) Requisite Skills** The ability to observe, recognize, and report problems to the AHJ.	2, 5	43–52

Chapter 6: First Responder Inspector (NFPA 1031) JPRs	Chapter(s)	Page(s)
6.5.6 Determine the operational readiness of existing fixed fire suppression systems, given test documentation and field observations, so that the systems are in an operational state, maintenance is documented, and deficiencies are identified, documented, and reported in accordance with the applicable codes and standards and the policies of the AHJ.		
6.5.6(A) Requisite Knowledge An understanding of the components and operation of fixed fire suppression systems and applicable codes and standards. **6.5.6(B) Requisite Skills** The ability to observe, make decisions, recognize problems, and read reports.	7	111–121, 123
6.5.7 Determine the operational readiness of existing fire detection and alarm systems, given observations, so that the systems are in an operational state, maintenance is documented, and deficiencies are identified, documented, and reported in accordance with the policies of the AHJ.		
6.5.7(A) Requisite Knowledge An understanding of the components and operation of fire detection and alarm systems and devices and applicable codes and standards. **6.5.7(B) Requisite Skills** The ability to observe, make decisions, recognize problems, and read reports.	6	84–99
6.5.8 Determine the operational readiness of existing portable fire extinguishers, given field observations so that the equipment is in an operational state, and deficiencies are identified, documented, and reported in accordance with the policies of the AHJ.		
6.5.8(A) Requisite Knowledge An understanding of portable fire extinguishers, including their components. **6.5.8(B) Requisite Skills** The ability to observe, make decisions, and recognize and report problems.	8	130–133, 135–137, 141–143
6.5.9 Inspect emergency access for an existing site, given field observations, so that the required access for emergency responders is maintained and deficiencies are identified and documented in accordance with the applicable codes and standards and the policies of the AHJ.		
6.5.9(A) Requisite Knowledge Policies of the AHJ, and emergency access and accessibility requirements. **6.5.9(B) Requisite Skills** The ability to identify the emergency access requirements and report deficiencies per the policies of the AHJ.	4, 9	58, 62, 149, 150
6.5.10 Recognize a hazardous fire growth potential in a building or space, given field observations, so that the hazardous conditions, material, liquids, or gases are identified, documented, and reported in accordance with the policies of the AHJ.		
6.5.10(A) Requisite Knowledge Fire behavior; flame spread and smoke development ratings of contents, interior finishes, building construction elements, decorations, decorative materials, and furnishings; and safe housekeeping practices. **6.5.10(B) Requisite Skills** The ability to observe, communicate, apply codes and standards, recognize hazardous conditions, and make decisions.	3, 4, 9	42–45, 60, 61–62, 148–159
6.6 Plans Review There are no plan review JPRs for first responder inspector.		

Glossary

A

Accelerator A device that accelerates the removal of the air from a dry-pipe or preaction sprinkler system.

Air sampling detector A system that captures a sample of air from a room or enclosed space and passes it through a smoke detection or gas analysis device.

Alarm initiation device An automatic or manually operated device in a fire alarm system that, when activated, causes the system to indicate an alarm condition.

Alarm matrix A chart showing what will happen with the fire alarm system when an initiating device is activated.

Alarm notification appliance An audible and/or visual device in a fire alarm system that makes occupants or other persons aware of an alarm condition.

Alarm valve A valve that signals an alarm when a sprinkler head is activated and prevents nuisance alarms caused by pressure variations.

Alternative clause This clause allows for the code provisions to be altered and an alternative offered that would not reduce the level of safety within the building.

Ambulatory healthcare occupancy A building or portion thereof used to provide services or treatment simultaneously to four or more patients that, on an outpatient basis. (NFPA 101, *Life Safety Code*)

Ammonium phosphate An extinguishing agent used in dry-chemical fire extinguishers that can be used on Class A, B, and C fires.

Anecdotal evidence Evidence that may or may not be true, used to generalize when there is insufficient data to base it upon.

Annealed The process of forming standard glass.

Annual inspections Inspections performed as part of the regular inspection cycle.

Apartment building is a building or portion thereof containing three or more dwelling units with independent cooking and bathroom facilities. (NFPA 5000)

Aqueous film-forming foam (AFFF) A water-based extinguishing agent used on Class B fires that forms a foam layer over the liquid and stops the production of flammable vapors.

Area of refuge An area that is either (1) a story in a building where the building is protected throughout by an approved, supervised automatic-sprinkler system and has not less than two accessible rooms or spaces separated from each other by smoke-resisting partitions; or (2) a space located in a path of travel leading to a public way that is protected from the effects of fire, either by means of separation from other spaces in the same building or by virtue of location, thereby permitting a delay in egress travel from any level. (NFPA 101)

Assembly occupancies Buildings (1) used for a gathering of 50 or more persons for deliberation, worship, entertainment, eating, drinking, amusement, awaiting transportation, or similar uses; or (2) used as a special amusement building regardless of occupant load. (NFPA 101, *Life Safety Code*)

Automatic sprinkler heads The working ends of a sprinkler system. They serve to activate the system and to apply water to the fire.

Automatic sprinkler system A system of pipes filled with water under pressure that discharges water immediately when a sprinkler head opens.

Auxiliary system A fire alarm system that sounds an alarm in the building and transmits a signal to the fire department via a public alarm box system.

B

Backdraft The sudden explosive ignition of fire gases when oxygen is introduced into a superheated space previously deprived of oxygen.

Beam detector A smoke detection device that projects a narrow beam of light across a large open area from a sending unit to a receiving unit. When the beam is interrupted by smoke, the receiver detects a reduction in light transmission and activates the fire alarm.

Bimetallic strip A device with components made from two distinct metals that respond differently to heat. When heated, the metals will bend or change shape.

Board of Appeals A group of persons appointed by the governing body of the jurisdiction adopting the code for the purpose of hearing and adjudicating differences of opinion between the authority having jurisdiction and the citizenry in the interpretation, application, and enforcement of the code. (NFPA 1)

Boiling liquid expanding vapor explosion (BLEVE) An explosion that occurs when a tank containing a volatile liquid is heated.

Business license or change of occupancy inspections Inspections that occur when the building department is notified of a new business requesting permission to open

Business occupancy An occupancy used for the transaction of business other than mercantile. (NFPA 101, *Life Safety Code*)

Butterfly valve A type of indicating valve that moves a piece of metal 90 degrees within the pipe and shows if the water supply is open or closed.

C

Carbon dioxide (CO_2) fire extinguisher A fire extinguisher that uses carbon dioxide gas as the extinguishing agent.

Carbon dioxide extinguishing system A fire suppression system that is designed to protect either a single room or series of rooms by flooding the area with carbon dioxide.

Central station An off-premises facility that monitors alarm systems and is responsible for notifying the fire department of an alarm. These facilities may be geographically located some distance from the protected building(s).

Check valve A valve that allows flow in one direction only. (NFPA 13R)

Chemical energy Energy that is created or released by the combination or decomposition of chemical compounds.

Chemical-pellet sprinkler head A sprinkler head activated by a chemical pellet that liquefies at a preset temperature.

Class A fires Fires involving ordinary combustible materials, such as wood, cloth, paper, rubber, and many plastics.

Class B fires Fires involving flammable and combustible liquids, oils, greases, tars, oil-based paints, lacquers, and flammable gases.

Class C fires Fires that involve energized electrical equipment, where the electrical conductivity of the extinguishing media is of importance.

Class D fires Fires involving combustible metals such as magnesium, titanium, zirconium, sodium, and potassium.

Class I standpipe A standpipe system designed for use by fire department personnel only. Each outlet should have a valve to control the flow of water and a 2½ in. male coupling for fire hose.

Class II standpipe A standpipe system designed for use by occupants of a building only. Each outlet is generally equipped with a length of 1½ in. single-jacket hose and a nozzle, which are preconnected to the system.

Class III standpipe A combination system that has features of both Class I and Class II standpipes.

Class K fires Fires involving combustible cooking media, such as vegetable oils, animal oils, and fats.

Clean agent A volatile or gaseous fire extinguishing agent that does not leave a residue when it evaporates. Also known as a halogenated agent.

Code A standard that is an extensive compilation of provisions covering broad subject matter or that is suitable for adoption into law independently of other codes and standards.

Code analysis A summary of the features of fire protection and building characteristics in a plan set.

Coded system A fire alarm system design that divides a building or facility into zones and has audible notification devices that can be used to identify the area where an alarm originated.

Combination smoke fire damper A device that functions as both a fire damper and as a smoke damper. (NFPA 5000)

Combustibility The property describing whether a material will burn and how quickly it will burn.

Combustible dust Any finely divided solid material that is 16.5 mil (420 μ) or smaller in diameter (material passing a U.S. No. 40 Standard Sieve) and presents a fire or explosion hazard when dispersed and ignited in air. (NFPA 654)

Commissioning The time period of plant testing and operation between initial operation and commercial operation.

Common path of travel The portion of exit access that must be traversed before two separate and distinct paths of travel to two exits are available. (NFPA 101, *Life Safety Code*) (NFPA 101)

Compartmentation The subdivision of a building into relatively small areas so that fire or smoke can be confined to the room or section in which it originates. (NFPA 232)

Complaint form Form that lists in detail any complaint that is lodged with the fire inspection agency and is investigated.

Complaint inspections Inspections that occur when someone registers a concern of a possible code violation.

Compressed gas cylinders Portable pressure vessels of 100 lb (45.3 kg) water capacity or less designed to contain a gas or liquid at gauge pressures over 40 psi (276 mPa).

Conduction Heat transfer to another body or within a body by direct contact.

Continuous beam A beam supported at three or more points. Structurally advantageous because if the span between two supports is overloaded, the rest of the beam assists in carrying the load.

Convection Heat transfer by circulation within a medium such as a gas or a liquid.

Corrosion-resistant sprinklers Sprinkler heads with special coating or plating such as wax or lead to use in potentially corrosive atmospheres.

Cylinders A portable compressed-gas container.

D

Day-care occupancy An occupancy in which four or more clients receive care, maintenance, and supervision, by other than their relatives or legal guardians, for less than 24 hours per day. (NFPA 101, *Life Safety Code*)

Dead end corridor A passageway from which there is only one means of egress. (NFPA 301)

Decay phase The phase of fire development in which the fire has consumed either the available fuel or oxygen and is starting to die down.

Defensible space An area as defined by the AHJ (typically a width of 30 ft [9.14 m] or more) between an improved property and a potential wildland fire where combustible materials and vegetation have been removed or modified to reduce the potential for fire on improved property spreading to wildland fuels or to provide a safe working area for firefighters protecting life and improved property from wildland fire. (NFPA 1144)

Deluge head A sprinkler head that has no release mechanism; the orifice is always open.

Deluge sprinkler system A sprinkler system in which all sprinkler heads are open. When an initiation device, such as a smoke detector or heat detector, is activated, the deluge valve opens and water discharges from all of the open sprinkler heads simultaneously.

Deluge valve A valve assembly designed to release water into a sprinkler system when an external initiation device is activated.

Detention and correctional occupancy An occupancy used to one or more persons under varied degrees of restraint or security where such occupants are mostly incapable of self-preservation because of security measures not under the occupant's control. (NFPA 101, *Life Safety Code*)

Dormitory A building or space in a building in which group sleeping accommodations are provided for more than 16 persons who are not members of the same family in one room, or a series of closely associated rooms, under joint occupancy and single management, with or without meals, but without individual cooking facilities. (NFPA 101, *Life Safety Code*)

Double-action pull-station A manual fire alarm activation device that requires two steps to activate the alarm. The person must push in a flap, lift a cover, or break a piece of glass before activating the alarm.

Dry chemical extinguishing system An automatic fire extinguishing system that discharges a dry chemical agent.

Dry sprinkler heads Sprinkler heads that are installed when 40 degrees Fahrenheit cannot be maintained. Dry sprinkler heads are constructed to provide isolation between the head and water supply by use of a cylinder that extends from the head to the threads of the pipe fitting. The threads reside in heated areas or are installed on a dry sprinkler system, and use pendent style dry sprinkler heads.

Dry-barrel hydrant A type of hydrant used in areas subject to freezing weather. The valve that allows water to flow into the hydrant is located underground and the barrel of the hydrant is normally dry.

Dry-chemical fire extinguisher An extinguisher that uses a mixture of finely divided solid particles to extinguish fires. The agent is usually sodium bicarbonate-, potassium bicarbonate-, or ammonium phosphate-based, with additives being included to provide resistance to packing and moisture absorption and to promote proper flow characteristics.

Dry-pipe sprinkler system A sprinkler system in which the pipes are normally filled with compressed air. When a sprinkler head is activated, it releases the air from the system, which opens a valve so the pipes can fill with water.

Dry-pipe valve The valve assembly on a dry sprinkler system that prevents water from entering the system until the air pressure is released.

Dry-powder extinguishing agent An extinguishing agent used in putting out Class D fires. The common dry- powder extinguishing agents include sodium chloride and graphite-based powders.

Duct detector A smoke detection device mounted either inside an HVAC duct or mounted on the outside of the duct with tubing arranged to sample the airflow to respond to the presence of smoke.

Dunnage Loose packing material (usually wood) protecting a ship's cargo from damage or movement during transport. (NFPA 1405)

Dust collection system A pneumatic conveying system that is specifically designed to capture dust and wood particulates at the point of generation, usually from multiple sources, and to convey the particulates to a point of consolidation. (NFPA 664)

E

Early-suppression fast-response (ESFR) sprinkler head A sprinkler head designed to react quickly and suppress a fire in its early stages.

Educational occupancies Buildings used for educational purposes through the twelfth grade by six or more persons for 4 or more hours per day or more than 12 hours a week. (NFPA 101, *Life Safety Code*)

Electrical energy Heat that is produced by electricity.

Electrostatic Particles or objects electrically charged with either a positive or negative voltage differential.

Enabling legislation Legislation in which local jurisdiction adopt a specific set of codes.

Endothermic Reactions that absorb heat or require heat to be added.

Equivalencies The use of systems, methods, or devices of equivalent or superior quality, strength, fire resistance, effectiveness, durability, to those prescribed by a code or standard.

Exhauster A device that accelerates the removal of the air from a dry-pipe or preaction sprinkler system.

Exigent circumstance An immediate life safety issue which requires that immediate actions be taken.

Exit That portion of a means of egress that is separated from all other spaces of a building or structure by construction or equipment as required to provide a protected way of travel to the exit discharge. (NFPA 101, *Life Safety Code*)

Exit access That portion of a means of egress that leads to an exit. (NFPA 101, *Life Safety Code*)

Exit discharge That portion of a means of egress between the termination of an exit and a public way. (NFPA 101, *Life Safety Code*)

Exothermic Reactions that result in the release of energy in the form of heat.

Extinguishing agent A material used to stop the combustion process. Extinguishing agents may include liquids, gases, dry-chemical compounds, and dry-powder compounds.

F

Film-forming fluoroprotein (FFFP) foam A water-based extinguishing agent used on Class B fires that forms a foam layer over the liquid and stops the production of flammable vapors.

Final notice Written correspondence used when violations are not corrected to notify the owner that legal action may be taken to ensure code compliance.

Fire A rapid oxidation process, which is a chemical reaction resulting in the evolution of light and heat in varying intensities.

Fire alarm control panel The component in a fire alarm system that controls the functions of the entire system.

Fire department connection (FDC) A fire hose connection through which the fire department can pump water into a sprinkler system or standpipe system.

Fire inspection A visual inspection of a building and its property to determine if the building complies with all pertinent statutes and regulations of the jurisdiction.

First responder inspector An individual at the first level of progression who has met the job performance requirements specified in this standard for first responder inspector. The first responder inspector conducts basic fire inspections and applies codes and standards. (NFPA 1030)

Fire inspector An individual at the first level of progression who has met the job performance requirements specified in this standard for fire inspector. The fire inspector conducts most types of inspections and interprets applicable codes and standards. (NFPA 1030)

Fire investigator An individual who has demonstrated the skills and knowledge necessary to conduct, coordinate, and complete an investigation. (NFPA 1033)

Fire load The weight of combustibles in a fire area or on a floor in buildings and structures, including either the contents or the building parts, or both.

Fire marshal A member of the fire department who inspects businesses and enforces laws that deal with public safety and fire codes.

Fire protection engineer A member of the fire department who works with building owners to ensure that their fire suppression and detection systems will meet code and function as needed.

Fire tetrahedron A geometric shape used to depict the four components required for a fire to occur: fuel, oxygen, heat, and chemical chain reactions.

Fire triangle A geometric shape used to depict the three components of which a fire is composed: fuel, oxygen, and heat.

Fixed-temperature heat detector A sensing device that responds when its operating element is heated to a predetermined temperature.

Flame detector A sensing device that detects the radiant energy emitted by a flame.

Flame point (fire point) The lowest temperature at which a substance releases enough vapors to ignite and sustain combustion.

Flameover (rollover) A condition in which unburned products of combustion from a fire have accumulated in the ceiling layer of gas to a sufficient concentration (i.e., at or above the lower flammable limit) such that they ignite momentarily.

Flammability limits (explosive limits) The upper and lower concentration limits (at a specified temperature and pressure) of a flammable gas or vapor in air that can be ignited, expressed as a percentage of the fuel by volume.

Flash point The minimum temperature at which a liquid or a solid releases sufficient vapor to form an ignitable mixture with the air.

Flashover A condition in which all combustibles in a room or confined space have been heated to the point at which they release vapors that will support combustion, causing all combustibles to ignite simultaneously.

Flow switch An electrical switch that is activated by water moving through a pipe in a sprinkler system.

Flush sprinkler A sprinkler in which all or part of the body, including the shank thread, is mounted above the lower plane of the ceiling. (NFPA 13)

Frangible-bulb sprinkler head A sprinkler head with a liquid-filled bulb. The sprinkler head activates when the liquid is heated and the glass bulb breaks.

Frequency How often a particular type of incident occurs.

Fuel All combustible materials. The actual material that is being consumed by a fire, allowing the fire to take place.

Fuel ladder A continuous progression of fuels that allows fire to move from brush to limbs to tree crowns or structures.

Fully developed phase The phase of fire development in which the fire is free-burning and consuming much of the fuel.

Fusible-link sprinkler head A sprinkler head with an activation mechanism that incorporates two pieces of metal held together by low-melting-point solder. When the solder melts, it releases the link and water begins to flow.

G

Gas detector A device that detects and/or measures the concentration of dangerous gases.

Gas One of the three phases of matter. A substance that will expand indefinitely and assume the shape of the container that holds it.

Growth phase The phase of fire development in which the fire is spreading beyond the point of origin and beginning to involve other fuels in the immediate area.

H

Halogenated extinguishing agent A liquefied gas extinguishing agent that puts out fires by chemically interrupting the combustion reaction between the fuel and oxygen.

Halon 1301 A liquefied gas-extinguishing agent that puts out a fire by chemically interrupting the combustion reaction between fuel and oxygen. Halon agents leave no residue.

Hardware The parts of a door or window that enable it to be locked or opened.

Hazardous material Any materials or substances that pose an unreasonable risk of damage or injury to persons, property, or the environment if not properly controlled during handling, storage, manufacture, processing, packaging, use and disposal, or transportation.

Healthcare occupancy An occupancy used for purposes of medical or other treatment or care of four or more persons where such occupants are mostly incapable of self-preservation due to age, physical or mental disability, or because of security measures not under the occupant's control. (NFPA 101, *Life Safety Code*)

Heat detector A fire alarm device that detects abnormally high temperature, an abnormally high rate-of-rise in temperature, or both.

Heat sync An object that, through conduction, draws heat away from a heat-producing object.

Hood and exhaust system Devices installed above a cooking appliance to direct and capture grease-laden vapors and exhaust gases.

Horizontal exit An exit between adjacent areas on the same deck that passes through an A-60 Class boundary that is contiguous from side shell to side shell or to other A-60 Class boundaries. (NFPA 301)

Hotel A building or group of buildings under the same management in which there are sleeping accommodations for more than 16 people and is primarily used by transients for lodging with or without meals. (NFPA 101, *Life Safety Code*)

Hydrostatic test A test filling the sprinkler piping with water and pressurizing it, usually to 200 psi (1379 kPa) for two hours, to look for leaks in the pipe work.

Hypoxia A state of inadequate oxygenation of the blood and tissue.

I

Ignition phase The phase of fire development in which the fire is limited to the immediate point of origin.

Ignition temperature The minimum temperature at which a fuel, when heated, will ignite in air and continue to burn.

Immediately Dangerous to Life or Health (IDLH) Any atmosphere that poses an immediate hazard to life or produces immediate irreversible debilitating effects on health. (NFPA 1670)

Incipient The initial stage of a fire.

Industrial occupancy An occupancy in which products are manufactured or in which processing, assembling, mixing, packaging, finishing, decorating, or repair operations are conducted. (NFPA 101, *Life Safety Code*)

Interior finish Any coating or veneer applied as a finish to a bulkhead, structural insulation, or overhead, including the visible finish, all intermediate materials, and all application materials and adhesives.

Intermediate level sprinklers Sprinklers equipped with integral shields to protect their operating elements from the discharge of sprinklers installed at higher elevations.

Ionization smoke detector A device containing a small amount of radioactive material that ionizes the air between two charged electrodes to sense the presence of smoke particles.

L

Large-drop sprinkler A sprinkler head that generates large drops of water of such size and velocity as to enable effective penetration of a high-velocity fire plume.

Line detector Wire or tubing that can be strung along the ceiling of large open areas to detect an increase in heat.

Liquid One of the three phases of matter. A nongaseous substance that is composed of molecules that move and flow freely and that assumes the shape of the container that holds it.

Loaded-stream fire extinguisher A water-based fire extinguisher that uses an alkali metal salt as a freezing-point depressant.

Lodging or rooming house Building or portion thereof that does not qualify as a one- or two-family dwelling, that provides sleeping accommodations for a total of 16 or fewer people on a transient or permanent basis, without personal care services, with or without meals, but without separate cooking facilities for individual occupants. (NFPA 101, *Life Safety Code*)

Lower explosive limit (LEL) The minimum amount of gaseous fuel that must be present in the air mixture for the mixture to be flammable or explosive.

M

Manual pull-station A device with a switch that either opens or closes a circuit, activating the fire alarm.

Master-coded alarm An alarm system in which audible notification devices can be used for multiple purposes, not just for the fire alarm.

Matter Made up of atoms and molecules.

Means of egress A continuous and unobstructed way of exit travel from any point in a building or structure to a public way, consisting of three separate and distinct parts: (a) the exit access, (b) the exit, and (c) the exit discharge. A means of egress comprises the vertical and horizontal travel and includes intervening room spaces, doorways, hallways, corridors, passageways, balconies, ramps, stairs, enclosures, lobbies, escalators, horizontal exits, courts, and yards. (NFPA 101)

Mechanical energy Heat that is created by friction.

Medical gases A patient medical gas or medical support gas. (NFPA 99)

Mercantile occupancy An occupancy used for the display and sale of merchandise. (NFPA 101, *Life Safety Code*)

Multipurpose dry-chemical fire extinguisher A fire extinguisher rated to fight Class A, B, and C fires.

N

Noncoded alarm An alarm system that provides no information at the alarm control panel indicating where the activated alarm is located.

Nozzle sprinkler head Sprinkler heads used in applications requiring special discharge patterns, such as directional spray or fine spray.

O

Obscuration rate A measure of the percentage of light transmission that is blocked between a sender and a receiver unit.

Occupancy The intended use of a building.

Occupant load The number of people who might occupy a given area.

One- or two-family dwelling A building that contains no more than two dwelling units with independent cooking and bathroom facilities. (NFPA 5000)

Open sprinkler heads A sprinkler that does not have actuators or heat-responsive elements. (NFPA 13)

Ornamental sprinklers Sprinkler that have been painted or plated by the manufacturer.

Outside stem and yoke (OS&Y) valve A sprinkler control valve with a valve stem that moves in and out as the valve is opened or closed.

P

Panic hardware A door-latching assembly incorporating a device that releases the latch upon the application of a force in the direction of egress travel. (NFPA 101B)

Pendant sprinkler head A sprinkler head designed to be mounted on the underside of sprinkler piping so that the water stream is directed down.

Performance-based code Outlines the requirement that a design has to meet, but does not state that a particular method or material must be used to meet the requirement.

Performance-based design A design process whose fire safety solutions are designed to achieve a specified goal for a specified use or application. (NFPA 914)

Permit A document issued by the authority having jurisdiction for the purpose of authorizing performance of a specified activity. (NFPA 1)

Photoelectric smoke detector A device to detect visible products of combustion using a light source and a photosensitive sensor.

Plume The column of hot gases, flames, and smoke that rises above a fire. Also called a convection column, thermal updraft, or thermal column.

Post indicator valve (PIV) A sprinkler control valve with an indicator that reads either open or shut depending on its position.

Preaction sprinkler system A dry sprinkler system that uses a deluge valve instead of a dry-pipe valve and requires activation of a secondary device before the pipes will fill with water.

Premise identification Posting of an address for emergency responders.

Prescriptive code Defines the specifics of a material of construction or action to be taken; such as the type of electrical wiring to use, based on the anticipated usage or requirement to conduct evacuation drills in a structure.

Proprietary supervising system A fire alarm system that transmits a signal to a monitoring location owned and operated by the facility's owner.

Protected premises fire alarm system A fire alarm system that sounds an alarm only in the building where it was activated. No signal is sent out of the building.

Pyrolysis The destructive distillation of organic compounds in an oxygen-free environment that converts the organic matter into gases, liquids, and char.

R

Radiation The combined process of emission, transmission, and absorption of energy traveling by electromagnetic wave propagation between a region of higher temperature and a region of lower temperature.

Rate-of-rise heat detector A fire detection device that responds when the temperature rises at a rate that exceeds a predetermined value.

Recessed sprinkler A sprinkler in which all or part of the body, other than the shank thread, is mounted within a recessed housing. (NFPA 13)

Reinspection An inspection performed to determine if code violations have been corrected.

Remote annunciator A secondary fire alarm control panel in a different location than the main alarm panel; it is usually located near the front door of a building.

Remote supervising station system A fire alarm system that sounds an alarm in the building and transmits a signal to the fire department or an off-premises monitoring location.

Residential board and care occupancy A building or portion thereof that is used for lodging and boarding of four or more residents, not related by blood or marriage to the owners or operators, for the purpose of providing personal care services. (NFPA 101, *Life Safety Code*)

Residential sprinkler system A sprinkler system designed to protect dwelling units.

Riser The vertical supply pipes in a sprinkler system. (NFPA 13)

S

Saponification The process of converting the fatty acids in cooking oils or fats to soap or foam.

Self-inspections Inspection performed by the building owner or occupant.

Severity The amount of death, injury, or damage that is the result of an incident.

Sidewall sprinklers A sprinkler that is mounted on a wall and discharges water horizontally into a room.

Simple beam Supported at two points neat its ends. In simple beam construction, the load is delivered to the two reaction points and the rest of the structure renders no assistance in an overload.

Single-action pull-station A manual fire alarm activation device that takes a single step—such as moving a lever, toggle, or handle—to activate the alarm.

Single-station smoke alarm A single device usually found in homes that detects visible and invisible products of combustion and sounds an alarm.

Smoke An airborne particulate product of incomplete combustion that is suspended in gases, vapors, or solid or liquid aerosols.

Smoke detector A device that detects smoke and sends a signal to a fire alarm control panel.

Solid One of the three phases of matter. A substance that has three dimensions and is firm in substance.

Spalling Chipping or pitting of concrete or masonry surfaces.

Spontaneous ignition Initiation of combustion of a material by an internal chemical or biological reaction that has produced sufficient heat to ignite the material. (NFPA 921)

Spot detector A single heat-detector device; these devices are often spaced throughout an area.

Spray booth A power-ventilated enclosure for a spray application operation or process that confines and limits the escape of the material being sprayed, including vapors, mists, dusts, and residues that are produced by the spraying operation and conducts or directs these materials to an exhaust system. (NFPA 33)

Sprinkler piping The network of piping in a sprinkler system that delivers water to the sprinkler heads.

Stakeholder An individual or group that is impacted by an issue.

Standard A document, the main text of which contains only mandatory provisions using the word "shall" to indicate requirements and that is in a form generally suitable for mandatory reference by another standard or code or for adoption into law. Nonmandatory provisions shall be located in an appendix or annex, footnote, or fineprint note and are not to be considered a part of the requirements of a standard.

Standpipe system A system of pipes and hose outlet valves used to deliver water to various parts of a building for fighting fires.

Stop work order A form used when contractors do not have the clearance for performing the work, or when work must be corrected prior to performing additional work.

Storage occupancy An occupancy used primarily for the storage or sheltering of goods, merchandise, products, vehicles, or animals. (NFPA 101, *Life Safety Code*)

Supervised Electronically monitoring the alarm system wiring for an open circuit.

Switch Any set of contacts that interrupts or controls current flow through an electrical circuit.

T

T tapping Improper wiring of an initiating device so that it is not supervised.

Tamper switch A switch on a sprinkler valve that transmits a signal to the fire alarm control panel if the normal position of the valve is changed.

Temporal-3 pattern A standard fire alarm audible signal for alerting occupants of a building.

Thermal column A cylindrical area above a fire in which heated air and gases rise and travel upward.

Thermal conductivity A property that describes how quickly a material will conduct heat.

Thermal layering The stratification (heat layers) that occurs in a room as a result of a fire.

Thermal radiation How heat transfers to other objects.

U

Underwriters Laboratories, Inc. (UL) The U.S. organization that tests and certifies that fire extinguishers (among many other products) meet established standards. The Canadian equivalent is Underwriters Laboratories of Canada (ULC).

Upright sprinkler head A sprinkler head designed to be installed on top of the supply piping; it is usually marked SSU ("standard spray upright").

V

Vapor density The weight of an airborne concentration (vapor or gas) as compared to an equal volume of dry air.

Variance A waiver allowing a condition that does not meet a recognized code or standard to continue to exist legally.

Vestibule A small room located between two spaces that provides an atmospheric separation for the purposes of controlling airflow or, in a smoke management system, the movement of contaminated air from one space to an adjacent space.

W

Wall post indicator valve (WPIV) A sprinkler control valve that is mounted on the outside wall of a building. The position of the indicator tells whether the valve is open or shut.

Water-motor gong An audible alarm notification device that is powered by water moving through the sprinkler system.

Wet-barrel hydrant A hydrant used in areas that are not susceptible to freezing. The barrel of the hydrant is normally filled with water.

Wet chemical extinguishing agent An extinguishing agent for Class K fires. It commonly consists of solutions of water and potassium acetate, potassium carbonate, potassium citrate, or any combination thereof.

Wet chemical extinguishing systems An extinguishing system that discharges a proprietary liquid extinguishing agent.

Wet chemical fire extinguisher A fire extinguisher for use on Class K fires that contains a wet chemical extinguishing agent.

Wet-pipe sprinkler system A sprinkler system in which the pipes are normally filled with water.

Wildland/urban interface Any area where wildland fuels threaten to ignite combustible homes and structures. (NFPA 1143)

Z

Zone of origin In the design of a smoke management system, refers to the smoke zone that the fire incident originates.

Zoned coded alarm A fire alarm system that indicates which zone was activated both on the alarm control panel and through a coded audio signal.

Zoned noncoded alarm A fire alarm system that indicates the activated zone on the alarm control panel.

Zoned system A fire alarm system design that divides a building or facility into zones so that the area where an alarm originated can be identified.

Index

Note: Page numbers followed by *f* and *t* indicate figures and tables, respectively.

T

U

V

W

Z